20 **Advances in Biochemical Engineering**

Managing Editor: A. Fiechter

Bioenergy

With Contributions by
H. L. Arora, H. W. Blanch, H. R. Bungay,
J. O. B. Carioca, M. M. Chang, L. F. Chen,
T. Y. C. Chou, Z. Duvnjak, K.-E. Eriksson,
M. C. Flickinger, Ch.-S. Gong, Th. Haltmeier,
A. S. Khan, N. Kosaric, M. Linko, B. Maiorella,
H. J. Potgieter, H. Sahm, G. G. Stewart,
G. T. Tsao, Ch. R. Wilke

With 58 Figures and 56 Tables

Springer-Verlag
Berlin Heidelberg GmbH 1981

ISBN 978-3-662-15366-6 ISBN 978-3-540-38553-0 (eBook)
DOI 10.1007/978-3-540-38553-0

Originally published by Springer-Verlag Berlin Heidelberg New York in 1981
Softcover reprint of the hardcover 1st edition 1981

Library of Congress Catalog Card Number 72-152360

2152/3020-543210

Preface

A. Fiechter, ETH-Hönggerberg, Zürich

Bioenergy obtained an enormous publicity during the past years. It was assumed as a real alternative to the replacement of fossil and nuclear energies. The initial enthusiasm was followed by a certain disillusionment, as besides the lack of economical technologies the feasibility of many well-meant propositions decreased drastically, due to the poor access to the annually synthesized 10^{11} t of biomass. The publication of volume 20 of our series coincides with a stage of development which shows, besides the setbacks, some realistic possibilities for the production of bioenergy from biomass.

This Jubilee Edition shall therefore be dedicated to this topic. A part of the largest integrated research program of the non-medical biology research area is placed into the foreground, e.g. the US program "Fuels from Biomass" with an annually granted sum of currently 100 million dollars.

The most prominent subject of these impressive R+D programs in many countries is the ethanol "gasohol" from sugar and high polymer carbohydrates (starch, cellulose) as well as methane from agricultural waste and sewage sludge. The wide scope of all the investigations is remarkable. This impression becomes evident when studying the reports from different countries also from outside the USA which are included in this selection of the American program. Many other countries which are not listed here have also started programs for gaining bioenergy. A complete list can hardly be made because of the dramatic development of the subject taking place very rapidly. The active scientists from the countries not quoted may forgive the editor for the lack of completeness due to the reasons mentioned. It is impressing that today the R+D for the development of biotechnical methods is highly promoted in all continents. In many cases, the final shape of the process design and the economy are not yet in sight and further efforts of biologists and engineers are required. It can be foreseen with certainty that today's work will result in an enormous support of biotechnology which will lead to significant reactions on biology and economy.

Despite the incompleteness of the selection of topics presented in this volume, it is hoped that the reader may obtain some characteristics of the present-time developments.

Each article has been prepared only recently. No reviewing has been done on them in order to preserve the new and original character of the writing and to allow the inclusion of the most recent results.

Undoubtedly, the reader will esteem the advantages of originality and topicality and overlook the disadvantages of incompleteness and minor insufficiencies in the finish of the writing.

Table of Contents

Biochemical Engineering for Fuel Production in the United States

Henry R. Bungay
Department of Chemical and Environmental Engineering,
Rensselaer Polytechnic Institute, Troy, N.Y., 12181 U.S.A.

Despite confusion, turmoil, and controversy, the United States is rapidly developing firm foundations for large scale production of fuels from biomass. Although methane is produced from manures at relatively large demonstration plants, this gains little attention compared to processes that lead to liquid fuels that could relieve the U.S. dependence on imported oil. Ethanol from corn grain is losing its political appeal, and other, cheaper feedstocks are being sought. There are several distinct processes being developed for lignocellulosic materials with a main difference being the type of pretreatment that permits good hydrolysis to degradable sugars. Byproduct credits and energy-efficient processes make it very likely that fuels from biomass will cost considerably less than imported oil, and it will be profitable to establish a massive new fuels industry.

1 Introduction

The use of energy in the U.S. has paused in its logarithmic growth because of some mild winters, more fuel-efficient automobiles, and conservation. The oil-exporting nations are curtailing production to reduce the glut resulting from the unexpected low demand. The production cost for oil is less than one percent of its selling price, so there is a danger that fuels from biomass could become uneconomic and producers could be bankrupted by a temporary lowering of the cartel's price of oil. Those entrepreneurs who produce fuels from biomass will be doing a great service to all the oil-importing nations of the world by introducing competition to oil, and they deserve protection against the unfair pricing practices of the cartel. Fortunately, there is excellent potential for obtaining fuels from processes which have coproducts that are valuable as food, fiber, or chemicals so that profits are possible even if fuel prices drop precipitously. High-priced ethanol from corn grain is sure to be replaced quickly by fuels from a new generation of biomass processes.

Burning of wood wastes to power lumbering and pulping operations, regional use of wood to heat home and buildings, and burning of bagasse contribute in excess of two quadrillion British Thermal Units (QUADS) of energy annually in the U.S. (1 QUAD = 2.52×10^{14} kilocalories.) This is not an insignificant percentage of the total annual consumption of energy of about 76 QUADS. Extrapolating from existing agricultural practices, it is very optimistic but not unreasonable for the U.S. to aim for 20 QUADS of biomass energy by the year 2000. If total energy usage does not resume its upward spiral, biomass can become a major contributor as the Earth's petroleum runs out.

The U.S. Department of Energy is relatively new but has a gigantic budget that increases each year. In its formative period there was a strong influence from the National Science Foundation because a number of people working on energy programs were transferred to the new agency then known as the Energy Research and Development Administration. Whereas the National Science Foundation has exemplary standards, the Department of Energy has evolved to a highly politicized system. There is some satisfaction in that almost all of the highly meritorious proposals are approved because the available monies are so plentiful. At present, ethanol from corn grain is losing some of its political glamour because the severe heat wave and drought of 1980 greatly elevated the price of corn. Cellulosic feedstocks are receiving attention just as some of the new processes are approaching commercial fruition.

Only recent work, most of it sponsored by the United States government, will be covered, and it is assumed that the reader appreciates the extensive previous research on cellulose hydrolysis and on ethanol formation throughout the world. Other chapters in this volume cover important candidate process or steps that are crucial to one or more processes. A review that ignored these other chapters would lack perspective, thus there is some duplication for purposes of comparison. The judgements expressed should not be taken as definitive because improvements and refinements are occurring at dizzying speed. This review will emphasize bioconversion and will have little to say about hydrogen from engineered photosynthesis. This fascinating, elegant approach will not reach fruition for many decades because there are many fundamental hurdles to overcome. One is the problem of generating hydrogen except at very low redox potentials, and photosynthetic evolution of oxygen nearby is troublesome to say the least. Other obstacles are devising cheap enclosures to capture gaseous products and separation of hydrogen from the other gases.

2 Organization for Biomass Research

The main divisions are:

production of land plants for feedstocks
production of aquatic plants for feedstocks
conversion to fuels by thermochemical processing
bioconversion
hydrogen from engineered photosynthesis

There is some biochemical engineering to growing plants, and culturing of algae demonstrates some sophisticated techniques. Space limits preclude topics other than bioconversion, but a detailed discussion of all the divisions is available [1]. Thermochemical processes are also excluded because there is little biochemical engineering.

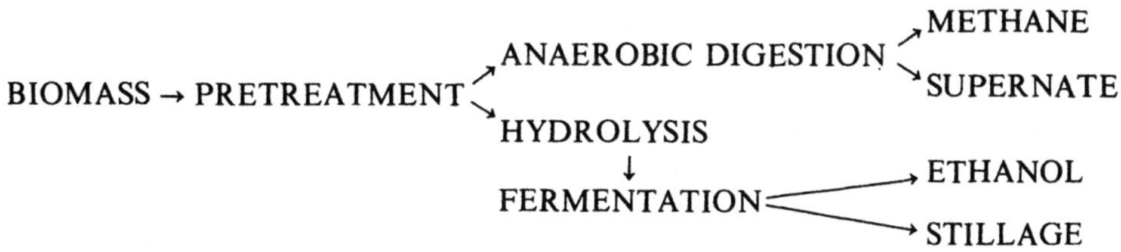

Fig. 1. Bioconversion processes

There is much enthusiasm for fuels from biomass by the thermochemical steps of gasification, pyrolysis, or liquefaction [2]. These are brute-force approaches that work with almost any carbonaceous material and produce simple organic molecules plus oils or tars. The economics of thermochemical conversion may be superior to those of existing bioconversion processes, but better fermentation yields and credits from byproducts are quickly reversing this situation.

The U.S. Department of Energy has had programs with various titles such as "alcohol fuels", "fuels from biomass", and "biomass refining", and administration is by D. O. E. itself or by other branches of government such as the U.S. Department of Agriculture. Portions of various programs are farmed out to branches of D. O. E. such as the Solar Energy Research Institute (SERI), Oak Ridge National Laboratory, and the like. There have been numerous temporary guidelines, but no long-range comprehensive plan for biomass fuels has been adopted.

The bioprocesses to be discussed are shown in Fig. 1.

Pretreatment can be omitted only for finely divided biomass such as tiny algae or loosely structured manures.

3 Anaerobic Digestion

Most work on anaerobic digestion features methane, but a few projects are aimed at recovering valuable organic acids. The great advantages for methane are: non-aseptic technique employing elective mixed cultures; production in relatively low-cost crude equipment; ease of recovering the insoluble product gas from water; and compatibility of the product with pipeline gas. Disadvantages are the low selling price of methane and $^1/_3$ to $^1/_2$ of the gas is carbon dioxide which lowers the fuel value. Only very inexpensive feedstocks can be digested economically. The demonstration plants and some commercial installations for manufacturing methane from biomass use cattle manure at very large feedlots. Much of the profit is derived by supplementing the cattle feed with digester residue that is high in protein. The value of the protein for refeeding is about twice that of the methane. The economies of scale are illustrated by the troubles of a skilled team of investigators who have been trying to show that a digestion operation at a small farm is an attractive investment [3,4]. However, several groups are perfecting packed anaerobic reactors which seem to have high efficiency [3,5].

All agricultural residues and trees are decomposed by anaerobic cultures, but yields vary [6]. Treatment of the biomass can improve digestibility, but it probably is not cost effective because of purchasing chemicals, equipment, operating the process, and introduction of sludges or salts from neutralization of the reagents [7]. Marine plants such as kelp digest well in saline systems or in fresh water after washing, but the cost of ocean-grown plants is presently far too high [8, 9]. Most of the organic matter in land plants is cellulose, hemicellulose, and lignin. While the carbohydrates digest rapidly and well, lignin is little effected. Some side chains are reacted or removed, but neither the ether linkages nor aromatic rings of lignin are attacked under anaerobic conditions.

Organic acids from anaerobic digestion may be recovered by a membrane process [10] or by solvent extraction [11]. The acids are valuable chemicals and could be used as fuels, especially if converted to hydrocarbons similar to gasoline.

Jones [12] has analyzed the economics of methane from anaerobic digestion of various materials and concluded that it is unprofitable unless the feedstock is priced well below the present costs of crops and agricultural residues, but others [13] are optimistic about digestion, particularly when manure is used.

4 Ethanol Processes

The most important processes for producing ethanol from biomass are shown in Table 1. The grain alcohol process is very popular presently because of the high subsidy provided by the Federal government and by several states for ethanol blended with gasoline. This program was intended to prop up the price of corn by creating more demand, but the drought of 1980 was of a serious nature and caused major price perturbations so that there is much less margin for profit at the prevailing price of corn in the Fall of 1980. Sugarcane juices and molasses are being processed to ethanol in Brazil at a large scale, and there are factories in other countries. With excess bagasses to fuel the factories and with low labor costs, the production of fuel alcohol is a good way to reduce requirements for imported oil. There have been several small technological advances, but the process relies on rather old technology. The wide distribution of cellulose and its relatively low price make it likely to become the main alcohol feedstock displacing corn and sugarcane.

The Natick process was the first significant advance in using cellulose to produce ethanol. Pretreatment by various types of grinding has proven too consumptive of energy. The molds which produce cellulase have been studied intensively by Reese, Mandels, and coworkers [15], and these efforts plus contributions of other groups (especially at Rutgers University) have led to excellent strains in terms of producing high titers of enzymes. The Berkeley process is derived from the Natick process and has contributed engineering solutions to most of the problems and has explored several alternative pretreatments. The economic prospects are good if uses can be developed for lignin and hemicellulose.

The Purdue group headed by Tsao showed great ingenuity in devising pretreatments and thus achieved nearly theoretical yields of glucose from cellulose. There are now several competing schemes at other institutions, but most resulted from the stimulus of the Purdue work. Other accomplishments are better dehydration methods for ethanol, various processes for the sugars from hemicellulose, different fermenter designs, and improvement of the solvent pretreatment to the point where good yields are

Table 1. Processes for manufacturing ethanol

Process	Description	Remarks
Grain alcohol	Corn grain is malted to hydrolyze the starch. Yeast produce ethanol and stillage is concentrated for cattle feed	Profitability can be destroyed by high corn prices or collapse of cattle feed market
Sugarcane	Juices or molasses are converted directly by yeast which are washed and recycled	Stillage too high in salts for cattle feeding. Credits for cane fiber could be high
Natick Process	Cellulosic materials treated with Trichoderma enzymes to get degradable sugars	Pretreatment by grinding too expensive. Has not focussed on using hemicellulose
Berkeley Process	Derived from Natick process and also uses hemicellulose	Strong candidate for large-scale operations
Purdue Process	Removal of cellulose and hemicellulose permits excellent hydrolysis with acid or enzymes	Regeneration of solvent may be costly, but this is a very high yielding process
Gulf Process	Enzymes added for simultaneous saccharification and fermentation	Hydrolysis yields not outstanding and good use of hemicellulose undeveloped
Pennsylvania/ General Electric	Solvent extraction of lignin gives excellent hydrolysis	Costly recovery of organic solvents
Iotech Process	Steam explosion fractures biomass for good hydrolysis	Very valuable lignin byproduct
M.I.T. Process	Mixed mold cultures hydrolyze biomass and produce ethanol	Simple but effective; highly promising

obtained by acid hydrolysis. Enzymatic hydrolysis is more expensive thus acid hydrolysis is presently featured at Purdue although yields are somewhat lower. Corn stover is probably the best cellulosic feedstock in the midwestern farm states.

The Gulf process appeared to be in the technological forefront just a few years ago, but newer processes have demonstrated superior yields. The concept of simultaneous hydrolysis and fermentation to relieve glucose inhibition of the hydrolysis of cellobiose has much merit, but the separate steps have different pH and temperature optima, thus process conditions require a compromise. Nevertheless, the simultaneous process deserves further research, and improvements such as better pretreatment of the biomass could revitalize its prospects.

A team effort of groups at the University of Pennsylvania and the General Electric Company has led to a process based on solvent extraction of lignin for better hydrolysis of cellulose and new thermophillic cultures to supply the cellulases. This is another highly promising process, and there are plans to get significant credits for byproduct lignin by such measures as using solutions in alcohols as diesel fuels.

The Iotech process uses steam explosion for pretreatment. High pressure steam permeates the biomass, and sudden release through a die shreds and disintegrates the structure. Hydrolysis of cellulose and conversion to ethanol proceed nicely. The biggest advantage, however, is development of high-value uses for lignin as a wood binder or specialty chemical. When there are many factories for fuel alcohol, the coproduct lignin will greatly overwhelm the foreseeable markets, but the first few factories selling lignin will be highly profitable. The search for new applications for lignin should be very rewarding because enormous quantities of material with superior properties compared to lignin from paper pulping will be available.

The M.I.T. process has more simultaneous steps than does the Gulf process. Carefully selected mixed cultures are added directly to coarsely ground biomass. Enzymes hydrolyze both the cellulose and the hemicellulose while the organisms convert the resulting sugars to ethanol. The organism which ferments the sugars from hemicellulose may be added later after the first organism has nearly completed the hydrolysis and has consumed most of the glucose. The really clever feature of this approach is investing very little in feedstock preparation and not being overly concerned with a high efficiency of feedstock utilization. This means that much of the feedstock is unreacted, but the residue does not represent much money. It would be burned to supply energy for the factory. Some improvement in efficiency of feedstock utilization would be desirable, however, because the fuel content of the residue exceeds the needs of the factory; steam or electricity would be products of about equal importance to the ethanol. There does not appear to be an opportunity to recover valuable lignin from the residue although it is enriched with respect to the other polymers. There are other problems such as inability of the present strains to reach high concentrations of ethanol, but the rate of accomplishment by the M.I.T. group has been outstanding.

Kelsey and Shafizadeh [14)] have still another simultaneous operation whereby the grinding of the feedstock is performed in the presence of cellulases. The rate of hydrolysis and the concentration of glucose were both improved.

5 Project Descriptions

Selected projects related to the U.S. program are shown in Table 2. Not all are currently active; some achieved their goals and were terminated while others are awaiting renewal of financial support before continuing. Those that are identified with processes listed in Table 1 are, of course, featuring further process development. Each project will be reviewed briefly.

The Natick group has performed excellent research despite rather erratic financial support. They selected grinding as a pretreatment step and other methods have proven superior. Lowest grinding cost results from wet milling between two rollers, but it is expensive compared to extraction or explosion techniques. As would be expected from reports of other groups, converting the Natick pilot plant to computer interfacing and control was very time consuming, but better operations and better analysis should repay the investment. Development of improved cultures for cellulase production has progressed well. Activity of beta-glucosidase is subject to biological controls different than those for cellulases, and optimum conditions are being defined. Pilot plant runs with *Aspergillus phoenicus* have shown good yields by either batch or continuous cultivation, thus supplementation with its beta-glucosidase should not be prohibitively expensive.

Engineering refinements of the Natick scheme and new departures are featured at Berkeley. A simple vacuum fermentation for ethanol has been superseded by a vacuum flash pot arrangement which allows escape of most of the carbon dioxide in

Table 2. Selected projects in the U.S. Biomass Program

Institution	Principal investigator	Description	Typical Reference
Natick Labs	Spano	Production of enzymes, hydrolysis ethanol formation	[15]
Berkeley	Wilke	Engineering aspects of complete process from cellulose to ethanol	
Pennsylvania/ General Electric	Pye	Complete process based on pretreatment by extraction of lignin	[17]
M.I.T.	Wang	Direct conversion of cellulose to ethanol. Several other processes	[18]
Purdue	Tsao	Complete process based on solvents for cellulose	[19]
Dartmouth	Grethlein	Kinetics of continuous acid treatment	[20]
Connecticut	Klei	Beta-glucosidase to improve cellulose hydrolysis	[21]
Auburn	Chambers	Extraction and fermentation of sugars from hemicellulose	[22]
Columbia	Gregor	Membrane steps in product recovery	[23]
N.Y.U.	Rugg	Acid hydrolysis in an extruder	[24]
Rutgers	Eveleigh	Genetics and selection of improved cultures	[25]
General Electric	Brooks	Pretreatment by chemically augmented steam explosion	[26]
Arkansas	Emert	Simultaneous hydrolysis and fermentation	[27]
Georgia Tech	O'Neil	Pilot plant for process comparisons	[28]
Battelle	Lipinsky	Sequenced process	[29]
Dynatech	Wise	Mixed-cultures for acids, then electrochemical conversion to hydrocarbons	[30]
Solar Energy Research Inst.	Villet	Complete process	[31]
Argonne Natl. Laboratory	Antonopolous	Cellulases from Fusarium	[32]
Mississippi State	McGinnis	Pretreatment by oxidation	[33]
Washington	Sarkanen	Solvent pulping with catalyst	[34]
Colorado State	Moreira	Organisms with alcohol tolerance	[35]

the main fermenter so that boiling in the flash pot gives a vapor rich in ethanol. Several different feedstocks have been tested for hydrolysis to sugars with enzymes from new and old strains of *Trichoderma reesei*.

Results have been good with a new process using hydrolysis with high pressure hydrogen chloride gas. Milled poplar wood dried to about 6% moisture absorbs the gas with a heat of solution. Yield is 75% reducing sugars. Although the yield is better with enzymes, acid hydrolyis is much cheaper and can be cost effective just as long as yields are reasonable. As with other acid hydrolyisis processes, the key is economical recovery and reuse of the acid. There is research on hydrochloric acid for hydrolysis at several institutions throughout the world, and acid recovery does not yet appear to have an economical answer.

The Berkeley group has also cooperated very effectively with other contractors and has integrated some efforts with the pilot plant at Gulf Oil Chemical Company at Pittsburg, Kansas. There have been economic evaluations of alternative processes and alternative feedstocks. One conclusion is that ethanol from cellulose is very likely to be uneconomic unless there are sizeable credits from products derived from hemicellulose and lignin.

A major realignment has taken place at the University of Pennsylvania because A. E. Humphrey, one of the preeminent biochemical engineers, has moved to Lehigh University. However, the logistics are good for cooperation between the universities and the General Electric Company. Work is also performed at Hahnemann Medical College in support of culture improvement. One similarity to the Berkeley group is the interest in vacuum fermentation. Thermophillic organisms permit reactions at a significantly higher temperature that gives a much higher vapor pressure thus milder vacuum means more attractive economics. Great attention has been paid to feedstock costs because they represent a major fraction of total cost, and tree nurseries in Pennsylvania have demonstrated high yields of poplar with short harvest cycles such that feedstock cost should be under $ 15 per dry ton. It is particularly interesting that the material produced from harvesting young trees two or three years old has about 20% of fines which can be separated easily by air classification. This fraction has 24–27% protein and an estimated price of $ 150 to $ 200 per t for animal feed.

Massachusetts Institute of Technology (M.I.T.) has one of the outstanding bioengineering groups in the world because of synergism of engineers, biochemists, geneticists, and microbiologists. Impressive results have come from developing microbial cultures with improved performance, as for example, higher titers of ethanol because of better ethanol tolerance. The old acetone/butanol process has been studied with more modern techniques such as pH control and nutrient feeding. Higher titers of butanol seem particularly difficult to reach because this alcohol is severely inhibitory at about $10\ g\ l^{-1}$. One promising approach is continuous removal of the butanol by extraction with a water-immiscible solvent during the reaction. Another example of innovative research is finding the bases for a commercial process to produce acrylic acid which is an important intermediate in manufacturing plastics and resins [35].

Current preoccupation with alcohol fuels has curtailed most of the financial support from D.O.E. except for the M.I.T. direct conversion of cellulosic materials to ethanol. Strains of clostridia have been developed that are approaching economically practical ethanol concentrations. Whereas the parent strains reached only a few grams per liter of ethanol, the selected strains achieve about $40\ g\ l^{-1}$. The parents had considerable lactic acid and roughly the same amount of acetate and ethanol, but the new strains have little lactate and 10 times as much ethanol as acetate. The best strain for hydrolysis of both cellulose and hemicellulose can degrade the resulting glucose to ethanol but does not use the sugars from hemicellulose. However, a second strain of clostridium does a fairly good job of degrading mixed sugars to ethanol. A brief review cannot do justice to the broad range of investigations by this group.

Other chapters in this volume describe the achievements of the Purdue group in defining the mechanisms of cellulose hydrolysis and the effects of various pretreatments. There has been both excellent theoretical research and practical engineering.

Much of their financial support has come from the State of Indiana which would like to assume a leadership role in alcohol fuels. The Purdue process employing solvents for destroying the crystallinity of cellulose can achieve yields approaching theoretical; the main problem is economical recovery and reuse of the solvents. This process can be characterized as high technology compared to the M.I.T. process which could be a very simple technology. It will be interesting to follow the various processes to see which becomes most successful.

The Dartmouth group has devised a continuous reactor for acid hydrolysis and determination of kinetic coefficients for the reactions occurring during hydrolysis and decomposition has been useful to all the other groups interested in acid hydrolysis. The small reactor used for several years has been replaced by a larger unit that permits higher pressures and provides data of greater reliability for scale-up. A particularly important discovery was pretreament by slightly less severe conditions of acid strength and temperature to remove hemicellulose hydrolytically and enable cellulases to achieve high yields of glucose in a subsequent step.

Some organisms produce a mix of cellulase enzyme activities that is deficient in beta-glucosidase. This enzyme converts cellobiose to glucose, but the rate declines as glucose accumulates. To maintain a high rate, excess beta-glucosidase is desirable. Several groups in the U.S. and other countries have shown that beta-glucosidase can be immobilized and used to supplement cellulase. Without immobilization, the cost is too high because the expensive enzyme is lost with each batch. A project at the University of Connecticut has found a good method for immobilization, and systems with the immobilized enzyme in an external column or right in the hydrolysis reactor have been demonstrated.

The group at Auburn University features hydrolysis of hemicellulose and fermentation of the resulting sugars. Red oak hardwood which is available locally has been hydrolyzed under a variety of temperatures and acid strengths to determine the kinetic coefficients. The maximum yield of xylose was 83% of theoretical when wood chips were treated; about 8% of the chip weight appeared as furfural. A trickling reactor minimizes dilution and produces fairly concentrated sugar solutions. The fermentation is also carried out in a packed reactor. Raschig rings or wood shavings are used to support and retain the microbial culture. Several different processes are being investigated, and results are encouraging for butanol production with little acetone and for the butanediol production.

Gregor at Columbia University has produced semipermeable membranes of a charged polymeric structure which resists fouling. Samples from projects in the U.S. and other countries have been subjected to various membrane treatments. Results with stillage from grain alcohol or from sugarcane alcohol have been excellent in terms of high flux rates, long membrane life, and low projected costs. Reverse osmosis instead of evaporation to concentrate stillage would reduce operating costs and also produce a more favorable overall energy balance. Recovery of ethanol by distillation is very easy in the range of the liquid-vapor equilibrium diagram where the curves are widely separated but difficult as the curves pinch together. It is attractive to distill to about 85% ethanol, and to complete the dehydration by a different method. Whereas the Purdue group is studying drying agents, the Columbia group is devising a membrane process based on polymers which have large flux differences for water and ethanol. It is too early to comment on the possibilities for success.

Another membrane process could have great significance throughout the process industries for recovery of acids and bases from salts. When a cationic and anionic membrane are close together and in a battery of alternating charged membranes, water can be split to hydrogen and hydroxyl ions which migrate to partner ions in the adjacent compartment. This means that feeding a salt such as sodium acetate to the water-splitting device will produce sodium hydroxide and acetic acid. Energy requirements are relatively low compared to electrolysis because ions are merely separated and not oxidized or reduced. Preliminary results at Columbia are highly encouraging.

A twin-screw extruder is used at New York University (N.Y.U.) to compact wood in sawdust form to a porous plug which is injected with dilute sulfuric acid at high temperature for hydrolysis. The hydrolysis rates agree well with kinetic coefficients reported at Dartmouth under similar conditions. However, with no pretreament, yields for this type of acid hydrolysis do not exceed about 55% of theoretical. The extruder has distinct sections, and it would be of great interest to use one or more sections for pretreatment. Another potential drawback is the expenditure of relatively large amounts of mechanical energy.

Some of the tedium of strain selection has been eliminated at Rutgers University by devising ways to accentuate the detection of colonies with improved properties on petri dishes. A number of strains have been found that produce high titers of cellulases, and the mix of activities is superior. Less inhibition of hydrolysis by the reaction products has also been achieved through mutation. A new project is improvement of Zymomonas strains; it is expected that genetic engineering will be easier with this bacterium that is a good producer of ethanol than with yeast.

A small unit for steam explosion of biomass has been operated at the General Electric Company. Addition of acids or bases increases the degree of pretreatment and improves the yield in the hydrolysis step.

The group led by Emert moved from Gulf Chemicals to the University of Arkansas. Research with better strains and process development continue.

Research at Georgia Institute of Technology may soon lead to a new process for inclusion in Table 1. Their original mission was to design and construct a pilot plant in which many of the alternative steps for converting biomass to ethanol could be compared. The current thinking at the Department of Energy is to build pilot plants dedicated to specific processes and not to perform side-by-side comparisons. The Georgia Tech group now has proposed a process based on steam explosion, acid hydrolysis in a screw extruder, and conventional steps thereafter. There has been little experimental verification of this scheme, but the concepts have merit.

One significant effort at Battelle/Columbus (Battelle Memorial Institute) has been promotion of sweet sorghum as an energy feedstock. This plant has many of the desirable features of sugarcane but has a wide geographic range encompassing most of the U.S. Biochemical engineering at Battelle has been overshadowed by the sweet sorghum project, but there has been work on a sequential system. The idea is to have a reaction step with a yeast well adapted to high sugar concentration in the first stage and another better adapted to high ethanol concentration in the second stage. This could be better than selecting a strain with compromised properties for both tasks.

Organic acids are key intermediates in the Dynatech scheme. The target is not ethanol, and the acids are to be converted by Kolbe electrolysis to hydrocarbons similar to components of gasoline. Anaerobic digestion with elective cultures is conducted in inexpensive reactors without aseptic techniques; it is well known that high loading with biomass gives good yields of acids. The seaweed, Irish Moss (*Chrondus crispus*), and fresh-water plants (water hyacinth, hydrilla, and duckweed) have been the featured feedstocks because the Dynatech group has been closely associated with research projects using aquatic biomass. However, other biomasses are also known to digest well. The acids are extracted continuously into kerosene and back into concentrated aqueous solution at higher pH. Feasibility has been established fairly well, and attractive economics have been forecast. Some others question using a premium energy form, electricity, to convert acids to cheap gasoline.

A broad range of solar energy projects including photovoltaic devices, windmills, solar boilers, and the like are missions of the Solar Energy Research Institute at Golden, Colorado. In the biomass area, much emphasis is given to the biochemistry and biochemical engineering of hydrogen from photosynthetic microorganisms. Some research on alcohol fuels from bioconversion is underway with strain selection, genetic modification, and process refinements.

Cellulases from *Fusarium* are being investigated at Argonne National Laboratories. The enzymes are known to possess good activity and stability, and the problem lies in the quite slow growth rate of these molds. *Fusarium* molds usually can use mixed sugars including those from hemicellulose, thus the implications are similar to the M.I.T. project where a superior culture can carry out several simultaneous operations.

Another type of pretreatment is being tried at Mississippi State University. Biomass is mildly oxidized in a wet state at elevated temperature and pressure. Yields in the hydrolysis step have not yet approached those of some using alternate pretreatment methods. Possible byproducts when oxygen is used are formic, acetic, and glycolic acid.

The group at Washington State University has many years of experience in wood pulping and hopes to exploit solvent pulping as a step in alcohol production. Ammonium sulfide is a pulping catalyst with aluminium salts in 50% ethanol for treating wood at 165 °C for 1 hour. This removes about 90% of the lignin, and the chips are easily disintegrated to low molecular weight cellulose fibers. The redirection to alcohol fuels is recent, and the potential of this approach cannot yet be assessed.

Inhibition of the culture by the alcohol product is the theme of research at Colorado State University. There is a search for tolerant strains resistant to either ethanol or butanol with emphasis on strains of *Clostridia* because these bacteria can use many sugars whereas the usual yeasts degrade glucose, fructose, and sucrose. A firm basis for selection is being laid by fundamental studies of the mechanisms of inhibition.

6 Recent Advances

Flickinger [37] has reviewed selected areas of research on degradation of cellulosic materials with emphasis on the present status and the potential for improvement.

In the brief time since this assessment, two groups have independently announced a remarkable improvement in fermentation of sugars from hemicellulose to ethanol [38, 39]. There are bacteria, molds, and yeast that degrade these sugars to ethanol, but other products are usually present and poor tolerance of ethanol prevents its accumulation. The best producers of ethanol are certain yeasts and the bacterium *Zymomonas*. Xylose, the predominant sugar from hemicellulose, is not used by the good ethanol producers, but xylulose, a keto sugar derived from xylose, is degraded well. When the enzyme glucose isomerase is added, xylose is isomerized to xylulose, but an equilibrium mixture that is still about 80% xylose is reached at prolonged times. This enzyme is widely used to convert glucose to fructose for commercial sweetners and it is inexpensive. A serious drawback is the need to recycle unreacted xylose back from the reaction step to the enzyme to again approach the concentrations of the equilibrium mixture. Work is underway to create mutants which have isomerase activity and thus need no supplemental enzyme. Furthermore, organisms which have the inherent ability to use xylose such as those being used at M.I.T. may soon be so improved that they merit commercial consideration. Utilizing hemicellulose to produce additional ethanol will mean a 50—60% improvement in productivity in factories using lignocellulosic biomass.

Other significant improvements are in fermenter design where there are several advantages to retaining organisms in the reactor or capturing them in the effluent for recycle. In addition to less diversion of substrates to growth, it is possible to overcome the inhibitory effects of alcohol on the microbial culture. There is a decrease in production rate on a per cell basis, but using massive numbers of cells restores the overall rate. Several new designs retain the cells to achieve very high populations [40]. One method uses heavily flocculated cultures which settle back as clear effluent is withdrawn from the top, and other designs have physical means such as immobilization or encapsulation to hold the cells in the fermenter. A group at Oak Ridge National Laboratory is having good success with *Zymomonas* held in a column reactor, and there is a good chance that this bacterium will outperform yeast in the future because plasmid transfer has been demonstrated in *Zymomomas mobilis* as a basis for genetic engineering [41].

Engineering problems are being solved by novel means for handling materials. Dilution is troublesome in several steps in the biomass processes because extraction yields are low unless excessive volumes of liquids are used. When biomass is mixed with water, the slurry concentration must be kept low or else stirring becomes impossible. Several groups are experimenting with contacting and extracting in columns with the liquid percolating through a solid bed. The solutions can be relatively concentrated so as to minimize the need for costly subsequent evaporation.

A number of economic estimates for producing ethanol from cellulosic biomass have appeared, but the crucial matter of coproducts has not been resolved. Using relatively low cost cellulosic wastes and taking credits for disposal of them leads to a favorable economic forecast [42]. However, projections are not so good when using crops, trees, or agricultural residues with value for competing uses [43]. Marginal economics will shift to a very attractive situation when ethanol yields are greatly improved by using sugars from hemicellulose and by selling lignin for a significant price.

7 Conclusion

Fractionation of biomass is leading rapidly to utilization of all its components. Hydrolysis of cellulose has improved in just a few years from yields in the range of 50% of theoretical to over 90%. Hemicellulose hydrolysis has always been easy, and there are highly promising ways for its conversion to ethanol. Lignin from the various biomass processes does not seem attractive for conversion by biological means, but it has great value in its native state because reactivity is much superior to lignin from paper pulping.

Methane is a logical product from biomass only when the feedstock is very, very cheap. As byproduct credits for using digester sludge as cattle feed are essential to process economics, there is a poor match of the energy product and the agricultural product to national needs. There is a similar criticism of matching byproduct lignin from ethanol factories to large markets, but the attractive features of reactive lignin present an exciting challenge for developing additional uses. Furthermore, lignin has more fuel value per unit weight than does ethanol and could be an energy supplement.

Individual steps in the processes for ethanol from biomass are achieving respectable yields. Some of the new processes have been carried out at bench or pilot plant scale all the way from feedstock to products; immediate construction of a large factory using the new techniques would be risky but not foolhardy. There is a great need for more research on recycle of nutrients from effluent streams back into the process, on low cost waste treatment or disposal of wastes by land irrigation, on energy integration of heat sources with heat sinks, and on yield improvements. Nevertheless, there are sound bases for predicting good economics for the new processes. The remaining hurdles for commercialization are not very difficult, and further refinements can change from good to outstanding profitability.

8 References

1. Bungay, H. R.: Energy, The Biomass Options. New York: Wiley 1981
2. Jones, J. L., Radding, S. B. (eds.): Thermal conversion of solid wastes and biomass: Am. Chem. Soc. Symposium Series No. 130 (1980)
3. Jewell, W. J., Dell'Orto, S., Fanfoni, K. J., Hayes, T. D., Leuschner, A. P., Sherman, D. F.: NTIS* SERI/TP-33-285, p. 547
4. Landers, T.: Proc. Bio-Energy '80, Bio-Energy Council, Washington, D.C., p. 138, 1980
5. Genung, R. K., Pitt, W. W.: NTIS SERI/TP-33-285, p. 437 (1979)
6. Pfeffer, J. T.: NTIS COO-2917-10 (1978)
7. McCarty, P. L., Young, L., Owen, W., Stuckey, D., Colberg, P. J.: NTIS SERI/TP-33-285, p. 411 (1979)
8. Dynatech R/D Company: NTIS HCP/ET-4000-78/1 (1978)
9. ibid: NTIS HCP/ET-4000-78/2 (1978)
10. Gregor, H. P.: NTIS SERI/TP-33-285, p. 39 (1979)
11. Sanderson, J. E., Wise, D. L., Augenstein, D. C.: Biotech. Bioeng. Symp. 8, 131 (1978)
12. Jones, J. L.: Chem. Eng. Prog. *76*, 58 (1980)
13. Klass, D. L.: Proc. Bio-Energy '80, Bio-Energy Council, Washington, D.C., p. 143, 1980

* N.T.I.S. is National Technical Information Service, U.S. Dept. of Commerce, Springfield, Va. 22161

14. Kelsey, R. G., Shafizadeh, F.: Biotech. Bioeng. *22*, 1025 (1980)
15. Ryu, D., Andreotti, M., Mandels, M., Gallo, B., Reese, E.: Biotech. Bioeng. *21*, 1887 (1979)
16. Maiorella, B., Wilke, C. R., Blanch, H. W.: Adv. Biochem. Eng. (this volume)
17. Ferchak, J. D., Hagerdal, B., Pye, E. K.: Biotech. Bioeng. *22*, 1527 (1980)
18. Wang, G. Y., Wang, D. I. C.: Abstracts of Papers, Am. Chem. Soc. National Meeting, Las Vegas (1980)
19. Gong, C. S., Chen, L. F., Flickinger, M. C., Tsao, G. T.: Adv. Biochem. Eng. (this volume)
20. Knappert, D., Grethlein, H., Converse, A.: Biotech. Bioeng. *22*, 1449 (1980)
21. Klei, H. E., Sundstrom, D. W., Coughlin, R. W., Ziolkowski, K., Biederman, G.: NTIS SERI/TP-33-285, p. 265 (1979)
22. Chambers, R. P., Lee, Y. Y., McCaskey, T. A.: NTIS SERI/TP-33-285, p. 255 (1979)
23. Gregor, H. P.: NTIS SERI/TP-33-285, p. 39 (1979)
24. Rugg, B., Brenner, W.: Proc. Bio-Energy '80, Bio-Energy Council, Washington, D.C., p. 160, 1980
25. Montenecourt, B. S., Schamhart, D. H. J., Cuskey, S. M., Eveleigh, D. E.: NTIS SERI/TP-33-285, p. 85 (1979)
26. Lamed, R., Su, T. M., Brennan, M. J.: Abstracts of Papers, Am. Chem. Soc. National Meeting, Las Vegas (1980)
27. Rivers, D. B., Emert, G. H.: Proc. Bio-Energy '80, Bio-Energy Council, Washington, D.C., p. 157, 1980
28. Roberts, R. S., Bery, M. K., Colcord, A. R., O'Neil, D. J., Sandhi, D. K.: Energy from Biomass and Waste IV, Inst. of Gas Technol., Chicago (1980)
29. Fink, D. J., Allen, B. R., Litchfield, J. H., Lipinsky, E. S.: NTIS BMI-2031 (1980)
30. Sanderson, J. E., Garcia-Martinez, D. V., Dillon, J. J., George, G. S., Wise, D. L.: NTIS SERI/TP-33-285, p. 97 (1979)
31. Villet, R.: Proc. Bio-Energy '80, Bio-Energy Council, Washington, D. C., p. 156, 1980
32. Antonopolous, A.: Report at Fermentation Contractors' Meeting, Sept. 1980: S.E.R.I. Alcohol Fuels R/D Newsletter, Golden, Colorado (in press)
33. McGinnis, G., Chen, C.: Report at Fermentation Contractors' Meeting, Sept. 1980: S.E.R.I. Alcohol Fuels R/D Newsletter, Golden, Colorado (in press)
34. Sarkanen, K. V.: Report at Fermentation Contractors' Meeting, Sept. 1980: S.E.R.I. Alcohol Fuels R/D Newsletter, Golden, Colorado (in press)
35. Moreira, A. R., Linden, J. C.: Report at Fermentation Contractors' Meeting, Sept. 1980: S.E.R.I. Alcohol Fuels R/D Newsletter, Golden, Colorado (in press)
36. Dalal, R. K., Akedo, M., Cooney, C. L., Sinskey, A. J.: Bioresources Digest *2*, 89 (1980)
37. Flickinger, M. C.: Biotech. Bioeng. *22*, Suppl. 1, 27 (1980)
38. Gong, C. S., Chen, L. F., Flickinger, M. C., Tsao, G. T.: Abstracts of Papers, Am. Chem. Soc. National Meeting, Las Vegas (1980)
39. Wang, P. V., Johnson, B. F., Schneider, H.: Biotechnol. Letters *2*, No. 6 (1980)
40. Sitton, O. C., Gaddy, J. L.: Biotech. Bioeng. *22*, 1735 (1980)
41. Skotnicki, M. L., Tribe, D. E., Rogers, P. L.: Appl. Env. Microbiol. *40*, 7 (1980)
42. Emert, G. H., Katzen, R., Frederickson, R. E., Kaupisch, K. F.: Chem. Eng. Prog. *76*, 47 (1980)
43. Fong, W. S., Jones, J. L., Semrau, K. T.: Chem. Eng. Prog. *76*, 39 (1980)

Structure, Pretreatment and Hydrolysis of Cellulose

Martin M. Chang, Terry Y. C. Chou* and George T. Tsao
Laboratory of Renewable Resources Engineering Purdue University
West Lafayette, IN 47907 U.S.A.

The structural features of cellulose which are important to enzymatic degradation are discussed. Pretreatments to facilitate the accessibility of cellulose are reviewed; in situ wet milling, semichemical pulping, and solvent pretreatment are considered the most effective to-date.

The hydrolysis of cellulose by enzyme is a complex phenomenon affected by both the structure of the substrate and condition of reaction. Analyses based on Michaelis-Menten kinetics had limited success. Results from traditional kinetic analysis indicated that cellulose was normally degraded

* Present Address: E. I. DU PONT DE NEMOURS & COMPANY, Experimental Station, Wilmington, Delaware 19898 U.S.A.

as if it were homogeneous for the enzyme. However, under some specific conditions, such as in situ wet milling and solvent pretreatment, a bimodal degradation might prevail. A comprehensive model to account for this variable mode of cellulose hydrolysis is discussed.

1 Introduction

Cellulose is the major component of cotton, wood, and biomass material. In fibrous form, it is the building material of all plant tissue. It can also be regarded as the energy captured by the green plant from the solar rays. As the fibrous and building material, cellulose has long been used by man as the most abundant and useful resource, but the importance of this role has been gradually reduced in recent years as cellulose products are replaced by products of modern technology, such as plastics and steel. Since the oil crisis of 1973, however, the use of cellulose as an energy source has gained renewed interest. Increasing research efforts have been directed to this area, particularly to the production of alcohol from the cellulosic biomass. Technically speaking, the production process is relatively simple — the hydrolysis of cellulose into glucose and subsequent conversion to ethanol. The process was used for large scale production during World War I and II and was further refined in the early 1940's [1]. Due to the low yield and corrosion by the catalytic acid, the process was not competitive economically with the petroleum-based processes. It has been gradually phased out in the postwar era. The recent energy crisis, however, has gradually reversed the situation. And the renewed research efforts may eventually improve the conversion process to the stage of economic reality. In this paper we will review the current state of knowledge about the structural features of cellulose and the means for breaking down this macromolecule with special emphasis on the enzymatic hydrolysis reaction.

2 Structure of Cellulosic Material

As the most abundant organic substance on the earth, cellulose is one of the thoroughly studied chemical compounds in science. But because of the limitation in instrumentation, several basic structural features of this substance are still unclear, namely, the direction or orientation of the long-chain molecule in the unit cell (parallel or antiparallel), the arrangement of the molecule in the crystallite (folded or extended chain) and the origin and mechanism of the morphologic transitions of the polymorphous cellulose. Since the hydrolysis of cellulose breaks down the molecule, the confusion about these features have made interpretation of the experimental observations difficult and, thus, might have hindered the progress of science in this relevant area. We will discuss these problems in some detail. Meanwhile, other features of cellulose essential to the hydrolytic reaction will be presented.

2.1 Crystalline Structures of Cellulose

2.1.1 Unit Cell Structure

Cellulose is a linear polymer of pure anhydroglucose units connected by 1,4β-glucosidic bonds. In nature, it exists in a highly organized state known as the fibrous crystal.

Table 1. Historical Development Regarding Molecular Orientation in Cellulose Crystal

Year	Author	Cellulose	Molecular orientation	Ref.
1928	Mayer & Mark	Ramie	Parallel	3)
1943	Mayer & Misch	Ramie	Antiparallel	4)
1962	Warwicker	Cotton	Parallel	6)
1974	Sarko, Blackwell	Valonia	Parallel	7, 8)
1974	Atalla	Reg. cellulose-I	Antiparallel	9)
1976	Blackwell, Sarko	Fortisan[a]	Antiparallel	10, 11)
1978	Chanzy	Shish kebab	Antiparallel	12)
1978	French	Ramie	Antiparallel	13)
1978	Chanzy	Single crystal	Antiparallel	14)
1978	Blackwell	Merc. cellulose	Antiparallel	15)

[a] Fortisan is a trade name of regenerated cellulose fiber

The basic repeating units of the crystal is the unit cell, which was defined by Meyer and Mark [2,3] as a monoclinic lattice with cellulose chains packed at the corners and the center of the cell. Later, the unit cell was redefined and named the "Meyer and Misch unit cell" of cellulose [4]. This structure has since been well received except for some minor corrections and some occasional disputes on the chain orientation. Table 1 sumarizes the most important developments to-date. A glance through the Table will show that the dispute has been a lengthy one, focussing mainly on whether the long-chain molecules should be oriented in a parallel or antiparallel formation. The confusion arises from the fact that cellulose is a paracrystalline substance but never is a perfect crystal. The X-ray diffractograph does not show sufficient data points to differentiate one formation from the other. The first model proposed by Meyer and Mark was a parallel formation [2]. This was later abandoned by Meyer for theoretical reasons but without experimental evidence. His reasoning was that the cellulose regenerated from solution should have antiparallel chains because this was the most convenient way for polymeric crystallization. The mercerized cellulose should also be antiparallel because its X-ray diffraction patterns were identical with those of the regenerated cellulose. As the latter was made from the native cellulose through a solid-state transformation during which a complete reversion of every neighboring chain was inconceivable if they were originally parallel, it could only be logical to assume that the antiparallel arrangement of the chains occurred in the native cellulose [5]. The development of the dispute since 1962 only repeats the internal argument of Meyer himself. The dispute is not yet completely settled and perhaps never will be, but the evidence strongly supports the antiparallel chain orientation in all cellulosic crystals except, perhaps, the *Valonia* cellulose.

Due to the complex pyranosyl structure, cellulosic unit cell may exist in a number of stable forms. Thus, there are five different lattice structures, the so-called Cellulose-I, -II, -III, -V, and -X [16]. By far, the most important structures industrially are Cellulose-I and -II. The former is the crystalline lattice for all native cellulose and the latter is the regenerated or mercerized form. The unit cell dimensions according to Meyer and Misch are a = 8.4 Å, b = 10.3 Å, c = 7.9 Å, $\theta = 84°$ for Cellulose-I and a = 8.1 Å, b = 10.3 Å, c = 9.1 Å, $\theta = 62°$ for Cellulose-II [4]. As a general rule, Cellulose-I is very easily converted to Cellulose-II. Treatment of cellulose, such as intracrystalline swelling, is sufficient to induce this conversion. But the reverse reaction is extremely difficult, if not impossible. There had never been any confirmed conversion until Atalla's experiment [9]. And in that work, the resultant Cellulose-I crystal could have been made of oligo-cellobioside instead of the macromolecular cellulose.

2.1.2 Crystallite Structure

The next structural feature of cellulose is the crystallite structure. Because cellulose is a long-chain molecule, the average molecular length ranges from 1,000 to 10,000 DP (degree of polymerization, in number average molecular weight unit; this is used throughout this paper except for literature data). In which way is the long chain oriented within the crystallite? This structural feature is far more important than many others of cellulose. Structurally, the smallest fibrous unit of cellulose is the elementary fibril, 35–40 Å wide and infinitely long [17]. These fibrils are in fact made of sequences of the so-called crystallite unit [18,19]. Figure 1 shows two representative models of the molecular orientation in the crystallite. The left-hand model is the fringed fibrillar model [20], which is a fibrillar version of the well known fringed micellar model. Accordingly, the cellulose molecules in the elementary fibril are fully extended with the molecular direction in line with the fibril axis. Among the fibril, however, there are intermittent highly ordered areas, the so-called crystalline regions, separated by less ordered or amorphous regions. The average length of the crystalline regions is about 500 Å for native cellulose and about 150 Å for the regenerated cellulose. These lengths have often been referred to as the leveling-off degree of polymerization or simply LODP because of the characteristic behavior of cellulose in the hydrolytic degradation [18]. Since the total length of the whole molecule is much longer than the LODP, a cellulose molecule normally will transit at least

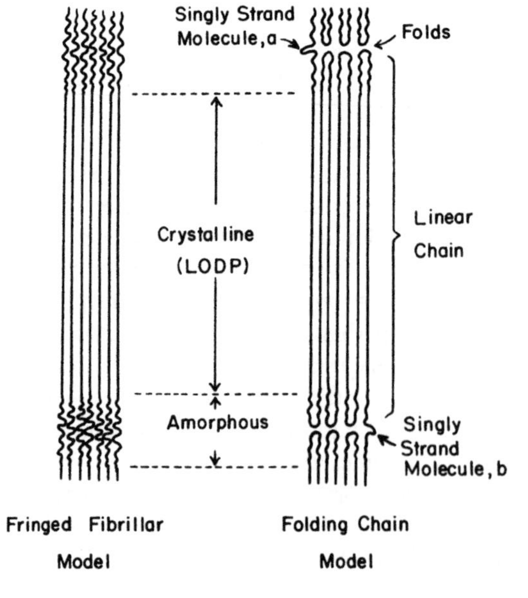

Fig. 1. Fringed fibrillar model and folding chain model of cellulose

10 consecutive pairs of crystalline and amorphous regions in a fibril. This is one of the basic molecular properties of cellulose.

The right-hand picture illustrates the folding chain model of the crystallite structure. Cellulose molecules here are visualized as being folded back and forth along the fibrillar axis within the 101 plane of the crystalline lattices. Thus, the folding molecule forms a sheet-like "platellite" unit at the fold length of approximately LODP. This makes up the basic molecular unit of the cellulose fiber [21]. A total of as much as 1,000 DP can be accommodated within this platellite unit (from "a" to "b" of Fig. 1). If the whole molecule is very much longer than 1,000 DP, the rest of the chain will enter into the neighboring platellite above or below in series along the elementary fibril. In this way, the corresponding portions of molecule connecting two platellites are single stranded chains and hang loose from the crystalline structure. These are the weak spots in the molecule vulnerable to relatively mild degradation, such as by exposure to light or by mechanical impact. But the breaking of these portions does not affect the physical and chemical properties of the cellulose fiber [21]. Another particularity of this model is that the glucosidic bonds at the folds (β_L-bond) are different. These are chemically much weaker than the linear β-bond and structurally very important to the integrity of the crystal. One bond of these broken per the basic molecular unit (1,000 DP for native and 300 DP for regenerated cellulose) will cause the disintegration of the crystal and severe loss in the mechanical strength of the fiber [21].

According to the folding chain model, the conventional crystallite is made of several platellites packed in crystallographic registry in the 101 direction. Along the longitudinal domain of the crystallite, there can be a transition of crystalline order. Thus, the amorphous regions are at the ends and the crystalline region at the center of the crystallite. Except for a few β_L- bonds at the folds, however, glucosidic linkages in both the crystalline and the amorphous regions are all linear β-bonds.

The folding chain model also suggests the multiple passages of the molecule through the amorphous and the crystalline regions like the fringed fibrillar model does. The differences between the two models are, first, a basic molecular unit of 1,000 DP is depicted by the former but not in the latter; second, a basic cellulose molecule will occupy much smaller space in the longitudinal direction (~500 Å) in the former than the latter (~5,000 Å); and third, there is a very small fraction of bonds (i.e., the β_L-bonds) that is structurally different from the other bonds in the so-called amorphous regions according to the folding chain model, while no such provision is given by the fringed fibrillar model. With these special molecular features, the folding chain model can provide better explanations of many indepth analyses of cellulose's properties than the fringed fibrillar model can [22].

The crystalline structure apparently plays a very important role in the hydrolytic degradation of cellulose. It has been observed that cellulose of high crystallinity reacted much slower than that of low crystallinity in enzymatic hydrolysis [23]. While the general observation is essentially correct, there can still be complications in the individual cases. We will elaborate further on this in the latter sections of this paper.

2.2 Anatomic Structure of Fiber

Because cellulose is a portion of the plant tissue, the structure of the plant may also play a role in the hydrolytic reaction. Anatomically, the wood cell is a multilayered structure. The first and external layer is called the primary wall. This primary wall is made of microfibrils oriented transversely to the long axis of the fiber. The next layer, the so-called "secondary wall", contains 3 sublayers — S1, S2, and S3. Each of these sublayers is formed during a particular growth stage of the cell and, hence, each contains structural units (i.e., microfibrils) with a different orientation that results in a laminated appearance. The most important of these sublayers is S2, which has the highest concentration of cellulose and makes up much of the secondary wall. Inside the secondary wall is the lumen, which is the portion of the living cell that contains the intracellular substance. When the cell is dead, this portion becomes a hollow space and is used by the plant to transport tissue fluid or water.

Within the fibrous tissue, the microfibril is a distinct structural unit, which contains much of the cellulosic material in the various layers and sublayers of the cell wall. The dimensions of the microfibril are generally 120 Å in cross-section and unlimited in length. Each microfibril contains several elementary fibrils bundled together laterally to give the distinct structure. The other components of the fiber, such as the hemicelluloses and lignin, are located between the microfibrils. Lignin is also found to be highly concentrated in the middle lamella or spaces between the fibers.

As a general rule, most of the cellulose is located in the S2 sublayer, where the microfibrils have a more orderly orientation than in the other layers. But this does not necessarily mean that the other layers are less important in cellulose hydrolysis. Because of the strategic positions of these outer layers, they can be the first to make contact with the catalyst and determine the pace of the subsequent reaction.

2.3 Lignin

Lignin is one of the three basic components of biomass. It is a three-dimensional polymer of phenolic origin with an infinite molecular weight. In the plant tissue, lignin functions as a preservative and as a cement between the individual fibers. Together with the hemicellulose, lignin matrix embraces the cellulosic microfibrils to form the protection sheath against foreign microorganisms. From chemistry point of view however, lignin may be neutral to the hydrolysis of cellulose. Once the cellulose is shaken loose from the lignin matrix by a suitable pretreatment, the hydrolysis proceeds smoothly and rapidly even though the lignin was apparently still present [24, 25].

2.4 Surface Structure

Since direct physical contact between the enzyme and the substrate is required for hydrolysis, the amount of surface area available to such contact is of primary importance to the reaction rate. In this respect, cellulose is a unique substance; it is not only a fibrous material with a very high surface-to-weight ratio but it is also a porous material with a potentially unlimited internal surfaces. Cowling and Brown described these two main types of surface as the gross capillaries (external surface) and the cell wall capillaries (internal surface) [26].

The external surface includes the gross surface of the fiber, the lumen, the pit apertures and pit-membrane pores that are visible under the light microscope. The total external surface area for cellulose fiber can be as great as 1 $m^2\ g^{-1}$. Since the cellulose fiber is much longer than wide, any size reduction does not increase the external surface area appreciably [27].

The internal surface of the fiber includes those spaces between the microfibrils and probably also the ends between the crystallites. Most of these internal surfaces are closed when the cell walls are free of water but open again when water is absorbed. The total internal surface area is a function of the pretreatment received by the fiber, such as pulping and beating. In general, this area is several orders of magnitude larger than the external surface. At full swelling, it may range from 300 to 600 $m^2\ g^{-1}$. The pore size of these capillaries also vary and can be as large as 200 Å, but many are smaller than 30 Å.

Due to the bulky size of the enzyme molecules, only a small fraction of these surfaces associated with the large pores are accessible to the enzyme. These are called the accessible surface to the enzyme. Stone and Scallan had examined this phenomenon and found a linear relationship between the initial reaction rate and the total surface that was penetrated by a probe molecule of 40 Å in diameter [27]. The experimental results leading to this conclusion is presented in Fig. 2. From these data, we can find that the maximum amount of surface available to the enzyme is about 150 $m^2\ g^{-1}$. Compared to the total internal surface area of cellulose fiber this is a relatively small fraction. It should be pointed out, however, that the surfaces in smaller pores are still there in the cell wall, but as far as the enzyme molecules are concerned, these surfaces are non-existent and therefore inaccessible.

3 Pretreatment of Cellulose

Due to the protection of the lignin sheath and the crystalline structure, the cellulose in wood or other plant tissue is normally not degradable by extracellular enzymes to any appreciable extent. To promote the saccharification a pretreatment of one kind or another is required. These can be divided into two major catagories: the physical and the chemical pretreatment. A number of different approaches have been tried but only those representative and highly promising are discussed here.

3.1 Physical Pretreatment

3.1.1 Ball Milling

The ball milling of cellulose may reduce the particle size, disrupt the crystalline structure and break down the chemical bonds of the long-chain molecule [28, 29]. With this much damage done to the cellulose, it is no wonder that the susceptibility was drastically increased [30]. But the energy consumption and therefore the cost of the process was formidably high; and the resultant overall rate of saccharification was

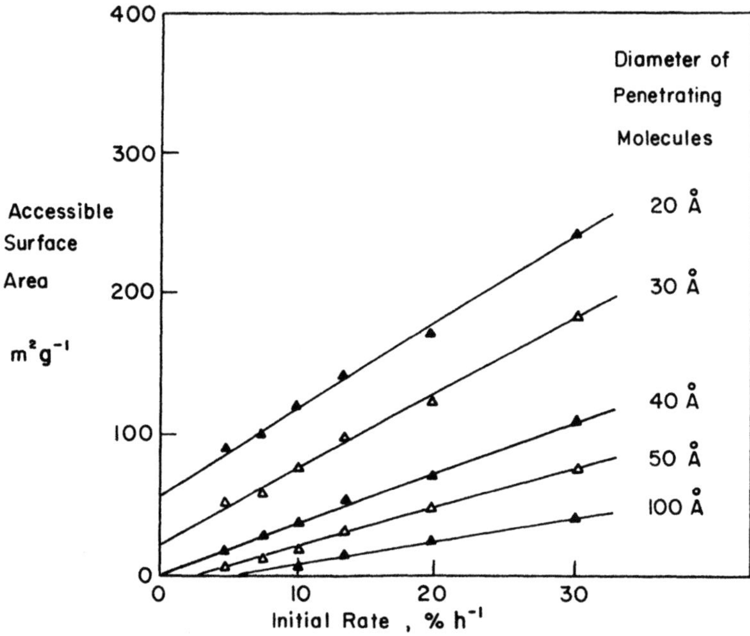

Fig. 2. Correlation between the rate constant and surface area of cellulose [27]

not adequate for industrial purposes. For instance, it took from 4 to 8 days for 90% conversion of cotton linters cellulose [28]. The cost of drying the substrate prior to the milling is another disadvantage.

3.1.2 Wet Milling

Wet milling or beating has been widely used in the pulp and paper industry to increase the surface area and the conformability of the cellulose fiber. But the application of this technique as a pretreatment for enzymatic hydrolysis was found either of no effect [27] or very much less effective than dry milling [20]. Wet milling can create the fibrillation and the delamination of cellulose [31], but the crystallinity and the chain length are unaffected, possibly due to the plasticizing action of the water [32].

While the wet milling is ineffective as an independent pretreatment stage, it can still be used to promote the enzymatic saccharification of cellulose. Kelsey and Shafizadeh have found very recently that the rate of hydrolysis could be increased substantially by a simultaneous wet milling and enzymatic degradation [33]. The corresponding increment in reactivity doubled that of the ball milling. The enzyme level they used was, however, impractically high at 40% of the weight of the cellulosic substrate.

In our study of the effect of solvent pretreatment of cellulose, we have also observed a profound increase in sugar yield by introducing a very mild grinding action into the reaction system [34, 35]. The mechanism of reaction was also changed from a pseudo first-order reaction to a bimodal degradation. For this reason we named this type of reaction the "mechano-enzymatic" hydrolysis. The interesting fact is that only a very mild mechanical action was actually needed to induce the effect. Thus, this type of approach seems to provide the economic feasibility needed for the overall saccharification process.

3.1.3 Steam-Explosion or Iotech Process

This process was invented by Delong and co-workers in Canada [36]. They employed the steam heating and rapid discharging technique to disrupt the woody texture for enzymatic accessibility. The basic process is very similar to the well known Masonite process [37]. Like the Masonite process, the wood chips were pressure-heated with steam in the digester to a temperature exceeding the softening point of lignin. Then the digester was abruptly opened to discharge the entire stock. Unlike the Masonite process, however, the details of the treatment condition and the end use of the product were different. This process produced a unique fibrous product in which the lignin no longer coated the carbohydrate components and could be easily extracted. The hemicellulose in wood was also partially depolymerized. An enormous amount of surface area was created, which might increase enzymatic accessibility.

3.1.4 Other Processes

Some other physical methods of activating the cellulose such as the high energy irradiation [38], heating in dry air or kerosene [39] and pressure treatment [40] were all found ineffective for enzymatic hydrolysis regardless of their cost. These will not be discussed here.

3.2 Chemical Pretreatment

3.2.1 Solvent Pretreatment

It has been well-established that the crystalline structure of the native cellulose can be completely destroyed by dissolving in a solvent, and that, upon reprecipitation, the cellulose is regenerated to a different but highly reactive form. The latter is the so-called cellulose-II crystal, to be differentiated from the native or cellulose-I structure. Welseth employed this principle in his study of enzymatic hydrolysis of cotton linters cellulose and observed a 10-fold increase in the extent of the conversion. The solvent he used was concentrated H_3PO_4 [40,41]. Although effective, the large quantity of the acid solvent that must be used makes the process uneconomical.

Ladisch and Tsao have found some other solvents of cellulose like Cadoxen to be very effective and proceeded to optimize the process through the recycling of the solvent [42]. To improve the economics further, a prehydrolysis was also introduced prior to the solvent pretreatment [43]. The overall process then included a low liquid ratio, dilute acid prehydrolysis to render the hemicellulose into simple sugars. The hydrolyzate was separated and treated further for conversion to alcohol [44]. The lignocellulose from the prehydrolysis was then treated by solvent to enhance the enzymatic accessibility. Cellulose solvents used for this step could be either Cadoxen or sulfuric acid. The capability of this type of pretreatment can be appreciated in Fig. 3. The saccharification is nearly completed in about 5 to 10 h. Under

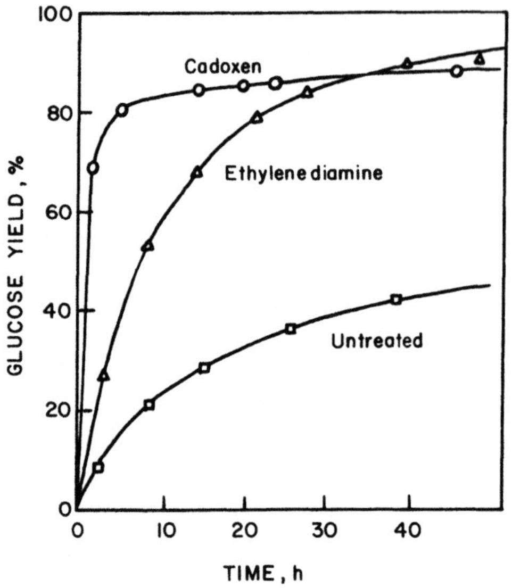

Fig. 3. Effect of solvent pretreatment of cellulose. Avicel cellulose was pretreated by Cadoxen solvent at room temperature, also presented is an ethylene diamine swollen Avicel cellulose and the untreated reference

the same hydrolytic condition, but without solvent pretreatment, only 40% of the untreated sample will be converted in a much longer reaction time (60 h). Interestingly, cellulose pretreated with a swelling agent — ethylenediamine — also shows a much higher conversion than the control although the reaction rate was slower than with the solvent pretreatment. On the other hand, the prescence of lignin in the pretreated substrate did not seem to interfere with the hydrolysis, lending support to the idea of the neutral role of lignin in the enzymatic reaction (see 2.3).

The key to industrial application of the solvent pretreatment process is the economy of recycling the solvent provided, of course, that the enzyme cost can be made equally competitive. Nevertheless, the enhancement in reactivity by the solvent pretreatment is not limited to the enzymatic reaction alone; the rate of the acid hydrolysis is also promoted and quantitative yield of sugar can be expected. If the solvent used was the concentrated sulfuric acid, the process would be essentially identical to that of Dunning with the exception of the solvent recycle [45]. Dunning's process was unsuccessful because of the high chemical cost, both for the acid and for the caustic used to subsequently neutralize the acid. Recycling the solvent appears to be a logical solution for this problem.

3.2.2 Swelling Pretreatment

There are two types of swelling for cellulose — "intercrystalline" and "intracrystalline". The first can be effected by water and is a prerequisite for any microorganism reaction to ever occur [46]. The second type of swelling requires a chemical reagent capable of breaking the H-bonding of cellulose. Aqueous solutions of relatively high concentrations of NaOH, organic bases (such as amines) and certain salts (such as $SnCl_4$) are reagents of this category. As a result of the intracrystalline swelling, the unit cell structure, the size of the crystallite and the crystallinity are all changed to produce an overall increase in reactivity. But this improvement was much less than that of the solvent pretreatment [47,48]. The high chemical consumption and low efficiency are the two major drawbacks of this approach.

On the other hand, the use of the swelling agent has been quite successful in the area of upgrading the nutritive value of the forage and forest residues for the animal feeds. This was accomplished at much lower level of chemicals, about 1–2% NaOH, than used for intracrystalline swelling [49]. Ruminant digestibility increased about 50% by a steeping treatment of straw in a 1.5% NaOH solution. Perhaps the difference here is that the microorganism was involved rather than an extracellular enzyme. The former is always superior to the latter in saccharifying power.

The application of heat in the presence of the swelling agent may offer an alternative to using large amounts of chemicals. A steeping treatment of bagasse with 2% NaOH at room temperature would probably not affect the enzymatic accessibility to any appreciable extent, but raising the temperature to 70 °C for 90 min was found to bring the sugar yield to a near completion according to Mandel and co-workers [50]. This treatment is essentially identical to the so-called semichemical pulping process. This type of pretreatment warrants further attention from researchers.

3.2.3 Pulping Process

Since the lignin sheath is a major deterent to enzyme reaction, any delignification process, such as the sulfite or kraft pulping, can be a potential candidate for pretreatment. But the conventional pulping processes are optimized to prevent the degradation of cellulose. This makes the process unnecessarily delicate and costly for our pretreatment purposes. As we pointed out previously, once the cellulose is shaken loose from the lignin, the mere presence of the latter does not interfere with the enzymatic reaction. So, a mild chemical processing to break up some lignin structure and to create sufficient surface area for rapid saccharification would be sufficient for our need. In this respect, the work of Baker and co-worker with SO_2 gas appeared to be promising [51]. The treatment involved a pressurized SO_2 cooking for 2–3 h at 120 °C. A nearly quantitative conversion to sugar was reported for hardwoods with this treatment. Slightly less conversion was observed for softwoods.

4 Hydrolysis of Cellulose

Hydrolysis of cellulose can be effectively catalyzed by both the acid and the cellulase enzyme. Due to the small molecular size, acid can penetrate deeply into the morphological structures of cellulose to effect a pseudo first-order sequential reaction [52]. The kinetics of the reaction has been thoroughly studied and accurately described according to Humphrey [53]. The relatively low yield, high by-product formation and high energy consumption are the major drawbacks of the process. Potentially, enzymatic hydrolysis may give a pure product at quantitative yield and consumes less energy. But the enzyme is a macromolecule. Its access to the heterogeneous and insoluble cellulosic substrate may be restrained by many factors not encountered by the acid catalyst. As a result, the reaction mechanism may be very different and, indeed, much involved. Due to the recent extensive pioneer work in enzymology and reaction kinetics, much in depth understanding of the reaction has been obtained and a large bank of useful information has been collected. However, many interpretations of the experimental observations were based on the fringed fibrillar model which might have been misleading due to the inaccuracy of the model. In the following review, we will discuss these observations in greater detail and offer an alternative interpretation from the folding chain model.

4.1 Current Understanding of Enzymatic Hydrolysis of Cellulose

4.1.1 Mode of Enzymatic Attack

Many excellent reviews on the mode of cellulase action have been published in the last few years [54, 55, 56, 57]. These can be summarized in the following. The action of enzyme leading to an effective breakdown of the native cellulose requires a sequentical operation of several basic cellulase components. These are the endoglucanase (hereafter symbolized by C_x), the exoglucanase (C_1) and the cellobiase (C_b). C_x is an endo- and random-cutting enzyme often referred to as CMCase. It is characterized by the release of free fibers from filter paper and the production of glucose from the carboxymethyl cellulose (CMC). The enzyme has been assumed to be capable of cutting the cellulose molecule in the middle of a chain to create reactive ends for the subsequent action of C_1. C_1, the exoglucanase, is an end-cutting enzyme with a nick-name "Avicellase". It is thought to be mainly responsible for the production of the cellobiose from the crystalline cellulose. C_b is an enzyme specific for cellobiose as the end product. Acting together, C_1 and C_x can solubilize native cellulose to cellobiose but neither C_1 nor C_x alone can do the job [58]. This is the so-called synergistic effect. There has been some different opinions about the action of C_1 [59], but in essence the synergistic actions of cellulase components have well been demonstrated [60]. This is the mode of action from the enzymologic point of view.

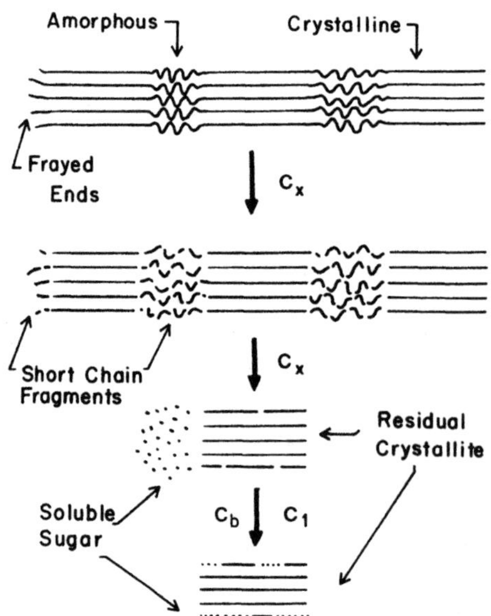

Fig. 4. The fringed fibrillar version of the degradation of cellulose by enzyme [54]

With respect to the structure of cellulose, it is generally regarded that the amorphous regions of cellulose is first hydrolyzed followed by hydrolysis of crystalline regions at a much slower rate [54, 56]. A well known picture of such a mode of degradation is reproduced here in Fig. 4. The fringed fibrillar model is used here to represent the original cellulose. Since this model implies that molecules in the amorphous regions are loosely bound, they should be easily reached and quickly degraded by enzyme. Thus the long chain molecules are broken first to the short crystallites, which are in turn further disintegrated, slowly perhaps, to simple sugars. In reality, this picture bears very little resemblance to experimental fact. This will be discussed in Sect. 4.2.1.

4.1.2 Kinetics of Hydrolysis

The kinetics of enzymatic hydrolysis has been intensively studied. Due to the diversity of the reaction system, numerous models were proposed. The major school of thought assumes the basic Michaelis-Menten kinetics.

$$E + S \underset{k_{-1}}{\overset{k_1}{\rightleftharpoons}} E \cdot S \xrightarrow{k_2} E + P \tag{1}$$

and

$$v = \frac{VS}{K_M + S} \tag{2}$$

where, E is the enzyme, $S = S_0 - P$, S_0 is the initial substrate concentration, P is the soluble product, [ES] is the enzyme-substrate complex, v is reaction rate, $V = k_2E_0$ or the maximum velocity, $K_M = (k_{-1} + k_{-2})(k_1)^{-1}$ or the Michaelis-Menten constant, and k_1, k_{-1}, k_2 are the rate constants.

The timecourse of sugar product can be derived as:

$$t = \frac{K_M}{V} \ln \frac{S_0}{S_0 - P} + \frac{P}{V} \tag{3}$$

A plot of P vs t gives the typical Michaelis-Menten plot as shown in the dotted line of Fig. 5. Also shown in this figure is a typical experimental timecourse of soluble sugar production for the enzymatic hydrolysis of cotton cellulose according to Mandel and Reese [61]. Comparing the experimental curve to the Michaelis-Menten, the rate of sugar production appears to slow down gradually during the reaction. A locigal interpretation to this deviation is the assumption of product inhibition to enzyme. Various types of inhibition had been introduced to modify the basic Michaelis-Menten equation [62, 63, 64]. For instance, Howell and Stuck assumed a non-competitive inhibition and derived the time-course in the following form [62]:

$$Vt = K_s\left(1 + \frac{S_0}{K_i}\right) \ln \frac{S_0}{S_0 - P} + \left(1 - \frac{K_s}{K_i}\right) P + \frac{P^2}{2K_i} \tag{4}$$

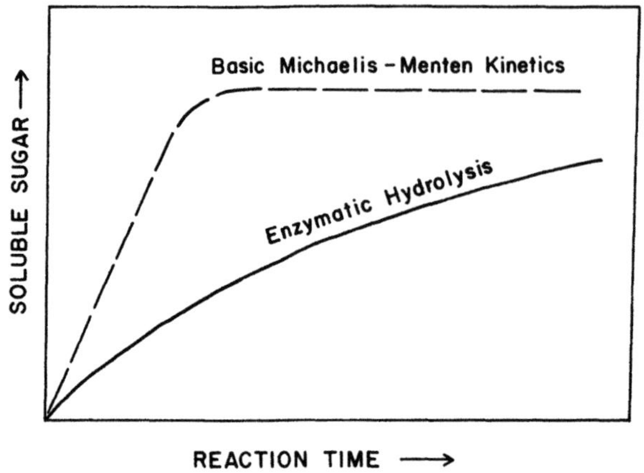

Fig. 5. Typical timecourses for the Michaelis-Menten kinetics and the experimental enzymatic hydrolysis of cotton cellulose

where K_s, K_i are the dissociation constants for the ES and EP complexes. Huang considered a fast adsorption of enzyme followed by a slow hydrolysis and subsequent product inhibition [63]. The timecourse he obtained also carried the form of:

$$t = \frac{1 + KE_0 + K'S_0}{k_2X_mKE_0} \ln \frac{S_0}{S_0 - P} + \frac{X_mK - K'}{k_2X_mKE_0} P \tag{5}$$

where K, K', k_2 are constants and X_m is the adsorption parameter. These and the other timecourse equations [64] can all be reduced to a general expression [62]:

$$t = A \ln \frac{S_0}{S_0 - P} + BP + DP^2 , \tag{6}$$

where A, B, D are the lump reaction constants related to K_M and other parameters of the enzyme-substrate-product system. For example, $A = \frac{K_M}{V}\left(1 + \frac{S_0}{K_i}\right)$, $B = \frac{1}{V}\left(1 - \frac{K_M}{K_i}\right)$, $D = \frac{1}{2V}\left(\frac{1}{K_c}\right)$ and K_i, and K_c are the inhibition constants to the enzyme and the complex of Eq. 1, respectively. With different combinations of K_i and K_c values, Eq. (6) may represent many types of product inhibition, such as $K_c = \infty$ for competitive inhibition, $K \neq K_c \neq \infty$ for the mixed type inhibition and $K_i = K_c = \infty$ for the basic Michaelis-Menten kinetics. Theoretically, the addition of the inhibition factor does conform the basic Michaelis-Menten plot to the proximity of the experimental curve. But, in reality, we found that the cited experimental curve in Fig. 5 happened to be exactly a simple and straight forward first-order timecourse, i.e., $t = A \ln [S_0/(S_0 - P)]$. If this were the case for cellulose hydrolysis, why should we go through the more complicated Michaelis-Menten mechanism at the first place?

We agree the Michaelis-Menten kinetics may be basically sound for many types of enzymatic reactions. But the direct application to the case of cellulose hydrolysis can be questioned for at least one good reason. That is, the cellulose is an insoluble substance. As such, the equilibrium condition required in Eq. (1) cannot be readily fulfilled. In this case the introduction of an inhibition factor to modify the case may not be the only realistic approach in resolving the intriguing problem. In fact, many of the above models exemplified by Eqs. (4) and (5) are applicable to some specific substrate only and are limited to reaction with a relatively low degree of conversion [62,63].

The traditional kinetic analyses of the enzymatic hydrolysis appeared to be more realistically founded. Ghose and Das proposed a pseudo first-order reaction pattern for the hydrolysis of the rice hull cellulose [65]. Their experimental data followed the proposed model very closely to about 50%–70% conversion of the potential sugar without the introduction of any product inhibition. Van Dyke pointed out that product inhibition is only a secondary factor and, instead, suggested a multiple component system for cellulose, all of which were first order to account for the changing reactivity during the course of the reaction [66]. Thus,

$$\frac{dP}{dt} = \Sigma\, k_i S_i\,, \tag{7}$$

where there were i different substrates (S_i), each with its own hydrolysis rate constant, k_i. Brant et al. obtained data from the hydrolysis of milled newsprint which lent very strong support to Van Dyke's model [67]. The inhibition factor was also ignored in their studies and yet the theory correlated very well with the experimental data.

Despite the success with the first-order kinetics by these workers, there were always some observations that did not obey the simple ruling, such as the swollen cellulose studied by Huang [63]. This was partly the reason for these many diversified studies on the subject. An overall comprehensive analysis of the system is perhaps appropriate at this time.

4.2 Outstanding Experimental Facts of Cellulose Hydrolysis

Among the numerous observations on the enzymatic hydrolysis of cellulose there is an outstanding feature which has not been fully recognized. This is — the cellulase may normally degrade the substrate molecule by entirety not by parts as has been suggested for acid hydrolysis. In another words, when cellulose is attacked by the enzyme, these long-chain molecules could possibly be distintegrated one molecule at a time without preferencial attack to any particular part of the molecule, such as the amorphous regions. This type of degradation pattern is entirely different from the conventional idea of cellulose hydrolysis (Fig. 4), which is that the long-chain molecule will be cut to shorter chains first and then further broken down to the small oligomers or soluble sugars. The experimental evidence leading to the former clue were scattered around in the literature so that no solid conclusions have been made on this issue. These will be presented in this section.

4.2.1 Consistent Molecular Weight

When native cellulose is degraded by enzyme, cellobiose and glucose are the two major products. Meanwhile, there is always some residual cellulose left depending on the extent of reaction. The molecular weight of this residue was found quite often

Table 2. Molecular weight measurement in enzymatic hydrolysis of cellulose

Year	Substrate	Enzyme	Molecular weight (DP)		
			Before hyd.	After hyd.	Ref.
1952	Cotton linters	Aspergillus	1385	1105	[41]
1957	Cotton	Trichoderma	4970	4200	[68]
1959	Wood pulp	Trichoderma	900	800	[69]
1965	Wood pulp	Penicillium	1150	400	[70]
1966	Cotton	Trichoderma	1320	1050	[71]
1967	Cotton	Basidiomycete	2260	1910	[72]
1967	Sulphite pulp	Trichoderma	No change in MWD[a]		[73]
1978	Cotton	Trichoderma	1800	1490	[74]

[a] Molecular weight distribution

to be the same as or only slightly reduced from the original samples. Table 2 summarizes the relevant experimental observations in the literature. Except for the case of the *Penicillium* cellulase, these data show quite clearly that cellulose molecules left after the enzymatic degradation are essentially intact. An overall average figure is less than one cut for every five molecules for these residual celluloses. The extent of hydrolysis ranges from 7% to 48% weight loss. Even for the case of the *Penicillium* cellulase the residual cellulose suffers only fewer than 2 broken bonds per molecule.

What these data really tell us is that at the molecular level the enzyme is essentially cutting a very small fraction of the substrate at a time. And, unless this fraction is completely degraded and dissolved into the solution, the enzyme will not attack some other molecules leaving the first undone. The mode of degradation fulfilling the above action should be a surface reaction instead of the random scission into the interior of the cellulose fibril as depicted in Fig. 4. The latter calls for a rapid drop in DP for the residual cellulose.

The above data also show another important feature of the cellulose property. Excluding the wood pulps, there is a definite molecular weight range centered at about 1000 DP, bayond which no degradation has gone. This behavior can be expected from the folding chain theory because this is the chain length accommodated in the basic crystalline structure "platellite" [21]. On this issue, the fringe fibrillar theory could not offer any logical explanation. Instead of 1000 DP, it would imply that 150 DP or the crystallite length would be the first stop of enzymatic degradation as represented by Fig. 4. We have just pointed out that this was not the case.

For swollen and regenerated cellulose, however, the situation can be quite different. The degradation did go beyond 1000 DP and then stopped at about 300 DP [41]. Due to the structural change imposed by the swelling treatment, the new platellite unit has a much shorter fold length (about 40 DP as compared to 150 DP for native cellulose). This new DP level of 300 is approximately the chain length enclosed within the "platellite" structure of the regenerated cellulose. The same clue about the consistent molecular structure may also hold for this type of cellulose.

Amemura's observation with *Penicillium* cellulase was the only other occasion in which the biological degradation of native cellulose had gone below the basic molecular level. The first observation of this kind was made by Cowling and Brown on the so-called brown root fungi, *Poria monticola*[26]. It could be possible that these microorganisms possessed some extraordinary capabilities for cutting the cellulosic chains, which should warrant further attention[22].

4.2.2 Consistent Crystallinity

An important behavior of cellulose can be anticipated from the observed invariant residual molecular weight, i.e., the crystallinity of the cellulose should also be invariant during the enzymatic process[22]. This statement directly contrasts with the common idea that "amorphous goes first" in cellulose hydrolysis[54,56,68]. On this issue, the direct experimental evidence in the literature have been controversial and inconclusive. Norkrans reported an increase in crystallinity for the enzymatic degradation of a reprecipitated cellulose[75]. But the X-ray measurement made by Cowling on samples of holocellulose showed a general decrease in crystallinity during the biological degradation process[76]. Other experiments did show some correlation between the crystallinity and the enzymatic digestibility of cellulose[23]. But these results could not be explained as any conclusive evidence which proved that the cellulase would preferably attack the amorphous regions rather than the crystalline regions, or vice versa.

Very recently, Bertrand and Buleon reported their work on the degradation of mercerized cellulose with *Trichoderma* cellulase[48]. A series of crystallinity determinations were concurrently made with measurements of the weight loss and chain length during the degradation process (Fig. 6). A small but significant increase in the crystallinity was noticed for the first 10% weight loss. From then on, the crystallinity remained essentially constant until about 70% of the cellulose was digested. Except for the initial increase in crystallinity, their observation lends very strong support to the anticipated consistent crystallinity for enzyme reaction.

The initial increment in the crystallinity deserves some explanation. Two possibilities could lead to the observed result. In the first, the crystallinity of cellulose in the primary

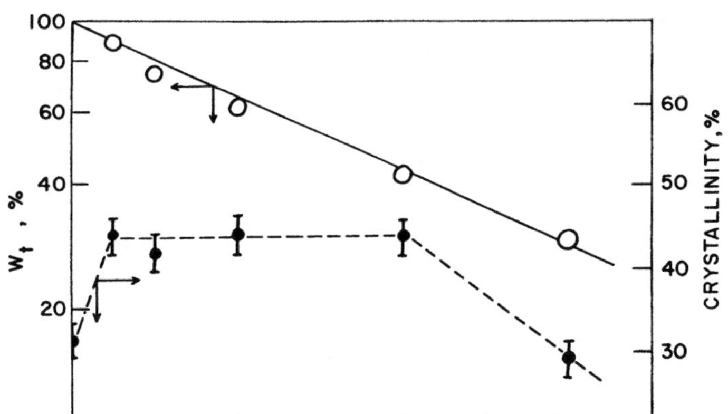

Fig. 6. Changes in crystallinity and residual weight of an alkali swollen cellulose in enzymatic hydrolysis[48]

cell wall and, perhaps, in the S1 and S3 layers might be lower than that of the bulk cellulose in the S2 layer. Due to their strategic positions these outer layers could be first digested by the enzyme to leave S2 cellulose behind. This gives an increase in crystallinity. The second possibility is that the cellulose sample may comprise more than one component with different crystallinity and reactivity. The component with higher crystallinity would have lower reactivity and vice versa. When enzymatic reaction is taking place, the first substrates to be digested should be those of high reactivity and low crystallinity which leaves the components of high crystallinity behind. Incidentally, the cause of the multiple components might come from a heterogeneous physical or chemical pretreatment that damages some portions of the fiber more heavily (such as fines or cracks) than others.

We have also made some crystallinity measurement on the residual cellulose from the enzymatic hydrolysis. The result essentially confirmed that of Bertrand and Buleon except for the initial changes in crystallinity, which we found to be either insignificant or nonexistent. This applied to both the native and the regenerated celluloses. But the partially swollen cellulose did behave differently. It initially showed a mixed crystalline pattern of cellulose-I and cellulose-II with low crystallinity. Upon progressive degradation, the cellulose-II pattern disappeared gradually while cellulose-I emerged. Eventually, there was only cellulose-I left with an increased overall crystallinity. Our results confirm both Norkran's and Bertrans' observations and are in harmony with the anticipated consistency theory. Details of these results will be presented elsewhere [77].

4.2.3 Quantum Mode of Degradation and Chain Orientation

Another important property deduced from the invariant molecular weight is the quantum mode of degradation. This is to say, when cellulose is attacked by enzyme, the macromolecule would be degraded directly from the molecular length down to cellobiose and glucose. There would be no appreciable amount of intermediate chain generated during the degradation process, at least not as a major product, which would otherwise affect the molecular weight reading of the residual cellulose. This statement is true for both the native cellulose and the hydrocellulose [68].

Reconciling this type of chain scission with the "degradation by entirety" derived in the previous Sect. 4.2.1, we can visualize an instant picture of cellulose molecule being degraded by the enzyme. The enzyme might start at the reducing end of the macromolecule and disintegrate the entire chain into cellobiose in one quick stroke. Since the chain was adherent to the fibrillar crystal to begin with, which is an immobile phase, the only logical way to accomplish the degradation is for the enzyme to move from one end of the modecule to the other. As we know the average length of the cellulose molecule is at least 5,000 Å and a typical enzyme is about 300–500 Å in length [26]. Such movement of enzyme is energetically inconceivable if the cellulose chain is fully extended in the crystalline state. But if the cellulose is folded, the entire molecular unit occupies a space of approximately the same size as the enzyme. The mode of enzymatic degradation should be more compatible with this type of molecular orientation that is with the folding chain model than with the extended chain model.

4.3 A Comprehensive Kinetic Model for Cellulose Hydrolysis

4.3.1 Hydrolysis under Normal Condition

As cited in the previous Sect. 4.1.2, Ghose and Das proposed apparent first-order kinetics for the enzymatic hydrolysis of cellulosic biomass [65]:

$$t = \frac{1}{k} \ln \frac{S_0}{S_0 - P} \tag{7}$$

A close examination of the literature data indicated that this system was applicable to many of the pure cellulosic substrates. A typical plot of weight-loss curves for cotton cellulose is shown in Fig. 7. These curves, taken from Selby's [78] and Reeses' data [61], do display simple first-oder kinetics throughout the entire course of the degradation. More interestingly, the extrapolations to zero reaction time pass exactly through the origin. This behavior suggests that in the "eyes" of the enzyme, the substrate is virtually homogeneous. As we know, one of the important experimental foundations for the two-phase theory — the amorphous and the crystalline regions of cellulose is the break in slope of the weight-loss curve [79] for the acid hydrolysis (see the dotted line in Fig. 7). In the absence of such a break, the two-phase theory loses its experimental base and the system should be regarded as a single phase [78]. This behavior matches perfectly the previously deduced theme of consistencies in both the chain length and crystallinity of cellulose. Thus, for the simplest case of cellulose degradation, the reaction scheme can be written as:

$$C \xrightarrow{k} P \tag{8}$$

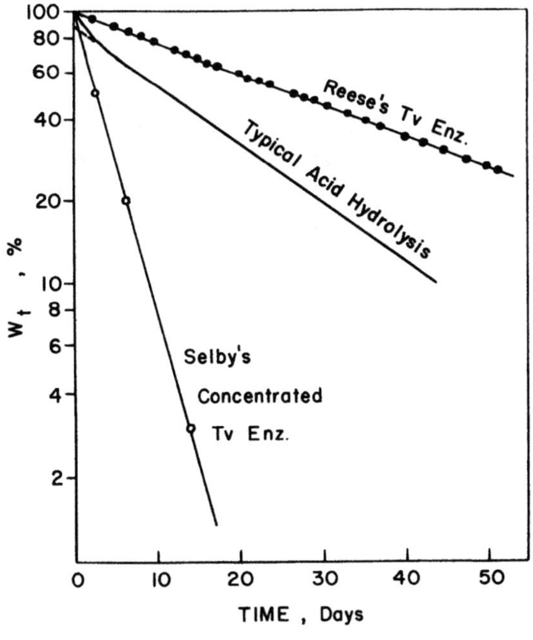

Fig. 7. Typical weight loss plots for the enzymatic hydrolysis of cotton cellulose and the acid hydrolysis

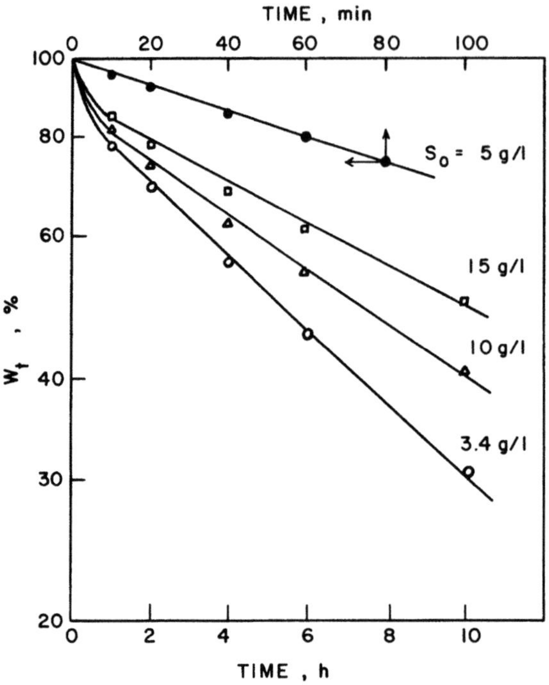

Fig. 8. Weight loss plots of enzymatic hydrolysis of Solka Floc cellulose showing behavior of multiple components

where C is the cellulose molecule at the original chain length, P is the soluble sugars including cellobiose and glucose, and k is the rate constant for the disintegration of the whole macromolecule.

While the cotton cellulose may behave as a simple homogeneous substrate to the enzyme, many of the industrial grade celluloses quite often show a two-component or multicomponent type of reaction (Fig. 8). The hydrolysis data of Howell and Stuck on the Solka floc cellulose [62] are typical. The lower three curves (in open points) do exhibit the two-components behavior typical of acid hydrolysis. The upper curve (solid circles) is a stretching of the initial period of hydrolysis. It is evident that both components for the initial period and for the subsequent reaction are first order but with different slopes. As postulated by Van Dyke [66], the degradation of the first and reactive portion of cellulose along with the others are all of first order. Therefore, a more general reaction scheme can be extended from Eq. (8).

$$C_i \xrightarrow{k_i} P_i \qquad (9)$$

where there are i different components, $C = \Sigma C_i$, each with its own rate constant, k_i. This can be regarded as the typical behavior of composite cellulose under normal incubation conditions.

4.3.2 Degradation of Cellulose under Shear

In the study of the effect of solvent pretreatment of cellulose, we detected a third type of enzymatic degradation in addition to the simple and the multiple first-order reactions [34,35]. Instead of the invariant molecular weight, a portion of the residual cellulose was found to be cut preferentially into an intermediate chain length characteristic of the crystallite structure of the substrate. A bimodal reaction scheme was proposed in the following form:

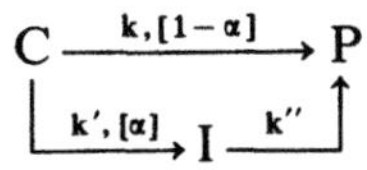

(10)

whereby α is the fraction of cellulose that undergoes the consecutive reaction, I is the intermediate chain at an average length of LODP of the substrate, k′ and k″ are the rate constants for the cutting of chain folds (β_L-bonds) and the linear chain (β-bonds) respectively. This type of reaction can be induced by either an in situ mechanical action, such as grinding, or a solvent pretreatment, which opens up the junctures between the crystallites in the elementary fibril [22]. Figure 9 reproduces the sequential changes in the molecular weight distribution for the residual cellulose during the mechanically induced biomodal degradation. Apparently this particular mode of degradation falls between the two extreme cases of cellulose hydrolysis — the acid and the enzymatic hydrolyses. In the former, C disappears very quickly and the I is the only species left in the residual cellulose. In the latter case, **I** never does appreciably appear and C is the only species. For the present case, both I and P coexist over a considerable length of reaction time. Since the in situ mechanical grinding is a necessary condition for at least the case of native cellulose, this mode of degradation is named the mechano-enzymatic hydrolysis [35].

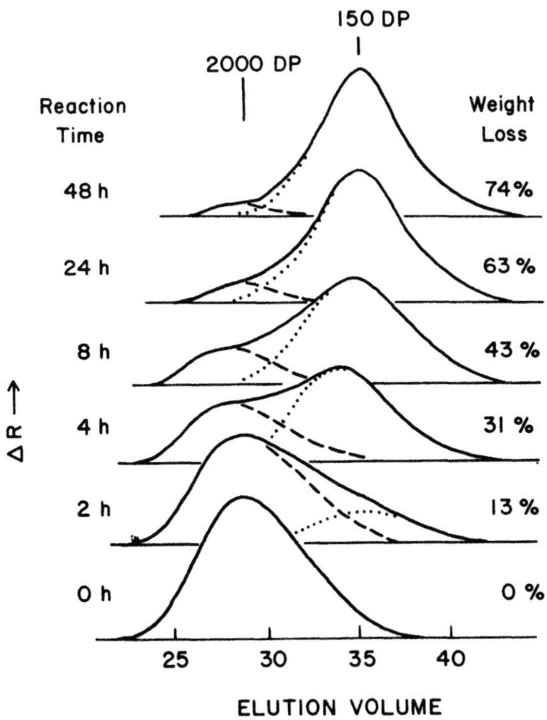

Fig. 9. Changes in molecular weight distribution of the residual cellulose during the mechano-enzymatic hydrolysis of cotton linters cellulose

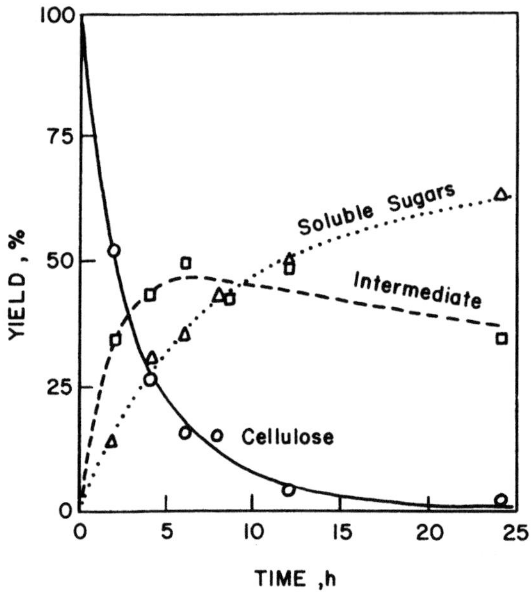

Fig. 10. Kinetic Analysis of the mechano-enzymatic hydrolysis of cotton linters cellulose. ○, □, △ data points for molecular cellulose, intermediate chains and soluble sugar; ———, ----------, ··········· predicted curves for molecular cellulose, intermediate chains and soluble sugars respectively

The kinetic behavior of the system is no longer obeying the first-order mechanism but conform to the bimodal consecutive reaction [34].
Hence:

$$C = S_0[\alpha\, e^{-k't} + (1 - \alpha)\, e^{-kt}] \tag{11}$$

$$I = S_0 \alpha k'(e^{-k'} - e^{-k''t})\,(k'' - k')^{-1} \tag{12}$$

$$P = S_0 - C - I\,. \tag{13}$$

For cotton linters cellulose, $\alpha = 0.51$, $k = 0.19\ h^{-1}$, $k' = 0.56\ h^{-1}$, $k'' = 0.014\ h^{-1}$. For the H_3PO_4 pretreated cellulose, $\alpha = 0.44$, $k = 0.72\ h^{-1}$, $k' = 6.6\ h^{-1}$, $k'' = 0.23\ h^{-1}$.

The timecourses for the above system, i.e., C, I, and P, for the hydrolysis of cotton linters cellulose are presented in Fig. 10. The predicted curves based on the bimodal reaction scheme do adequately match the experimental points without the introduction of the inhibition factor.

Up to the present point it is feasible to make a qualitative account of the enzymatic action (Fig. 11). Under normal circumstances, the cellulase appears to attack only a small fraction of the substrate molecules at a time and leaves the other fraction untouched. There seems to be no preferencial cut on the amorphous regions or chain folds of the cellulose molecule, as represented by Fig. 11 b. In some specific cases, such as when a mechanical force is acting on the cellulose molecule, some of the chain folds may be exposed to the cellulase to be cut faster than the linear chains are cut to leave the straight-chain segments at LODP. In this way, the so-called hydrocellulose is formed as the intermediate degradation product. Meanwhile, there are still some of the whole cellulose molecules left intact to co-exist with the newly formed

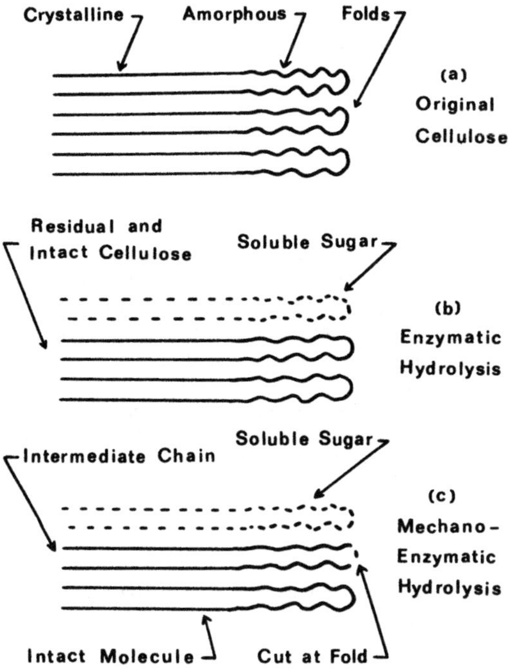

Fig. 11. Folding chain versions of enzymatic degradation of cellulose

hydrocellulose and, of course, the final products cellobiose and glucose (Fig. 11c). Because of these different responses, we can see that both the condition of reaction and the state of the substrate may affect the mode of cellulase action. In addition, the cellulase enzyme is itself a composite catalyst of different chain-cutting capabilities (C_x, C_1, C_b). It is quite clear that there is inevitably much complexity in the enzymatic degradation of cellulose.

4.3.3 A Comprehensive Model of Cellulose Hydrolysis

To describe the kinetic behavior of the partially swollen cellulose, Chou further extended Eq. (10) through (13) to accommodate the multiple-component system [34]. The GPC analysis indicated that the partially swollen cellulose was composed of a mixture of cellulose-I and -II. During the enzymatic degradation, each of these components was reduced to soluble sugar independently, through a process typical of the bimodal consecutive reaction described by Eq. (10). A general formula can be written as:

$$C_i \xrightarrow{k_i,[1-\alpha_i]} P_i, \qquad C_i \xrightarrow{k'_i,[\alpha_i]} (HC)_i \xrightarrow{k''_i} P_i \tag{14}$$

whereby C_i = cellulose component i, P_i = soluble sugar generated from component i, $(HC)_i$ = hydrocellulose generated from component i, α_i = fraction of component i that undergoes the consecutive reaction, k_i, k'_i and k''_i = rate constants of component i. The application of this model to the hydrolysis of partially swollen cotton linters

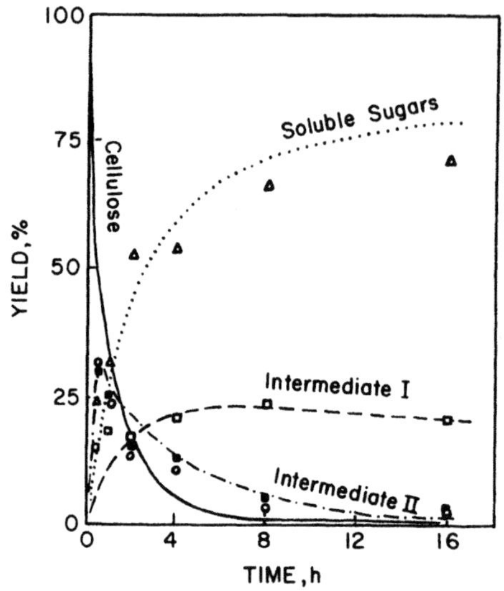

Fig. 12. Kinetic analysis of the mechano-enzymatic hydrolysis of partially swollen cotton linters cellulose. ○, □, ■, △, data points for molecular cellulose, intermediate chain of cellulose-I, intermediate chains of cellulose-II, and soluble sugars respectively; ————, -------------, —.—.—.—, ············ predicted curves for molecular cellulose, intermediate chain of cellulose-I, intermediate chain of cellulose-II and soluble sugars respectively

cellulose by H_3PO_4 was found to be basically satisfactory except that it over predicted the sugar yield by about 10% at high conversion (Fig. 12).

To summarize the development of the traditional kinetic analyses from Ghose to Chou, we can make a tentative conclusion on the multiple mode of cellulose hydrolysis. The overall degradation model can be represented by Eq. (14). For cellulose of a single component and under the normal reaction conditions, $i = 1$ and $\alpha_1 = 0$. The enzyme will attack the substrate as if it were homogeneous despite the two-phase crystalline structure. The degradation proceeds in the pseudo first-order mechanism. The residual weight of the cellulose then is a simple exponential function of the reaction as described by Ghose and Das [65].

Under some specific conditions, however, the weak bonds (β_L-) of cellulose can be preferentially accessible to the enzyme. The concurrent biomodal type of degradation may take place. Here $i = 1$ and $\alpha_1 \neq 0$. A fraction of the cellulose will undergo the consecutive reaction to generate some intermediate chains in addition to the direct disintegration. The reaction kinetics can be defined by Eqs. (11) through (13). An example of this type of degradation is the so-called mechano-enzymatic reaction.

For cellulose of multiple components, both i's and α's may vary depending on the individual case. Factors such as the history of the pretreatment and the conditions of the enzymatic reaction may all play a role in determining the overall reaction mechanism of the degradation processes. Equation (14) has been tested for the two-component system [34]. Other cellulosic substances, including conventional biomass, are being investigated in our laboratory.

5 Conclusions

Chemically speaking, cellulose is one of the simplest natural polymers, but, due to the insolubility of the substance, the complexity of the plant tissue, the flexibility of the crystalline structure and the lamella surfaces, the breaking down of this polymer by enzyme can be a very involved process. Many factors relating to the substrate as well as to the reaction conditions may affect the mode of enzymatic degradation. The situation is compounded by the multiplicity of enzyme components. Hence the detailed mechanism of the enzymatic hydrolysis of cellulose is yet to be fully understood.

Among the structural features of vital importance to the enzymatic accessibility of cellulose are the specific surface area available to the enzyme, the crystallinity level of cellulose and the state of lignin seal in the biomass matrix. As these features can be altered by either a physical or a chemical treatment, the choice of a suitable pretreatment to enhance the accessibility has become an important measure in the biomass conversion process. To this end, the in situ wet milling or the mechano-enzymatic technique showed some promise for the needed action. The semi-chemical pulping processes could be another alternative to improve the accessibility. The solvent pretreatment might also be practical if the solvent recycling stage could be made economical.

The hydrolysis of cellulose catalyzed by enzyme is a complex reaction. Up to the present theoretical kinetic analyses using the Michaelis-Menten mechanism supplemented by an inhibition factor has only limited success. The traditional approach so far has shown that, under the normal reaction conditions, cellulose is degraded by enzyme as if it were a homogeneous substrate. Both the molecular weight and the crystallinity of cellulose are essentially unchanged after the reaction: The long-chain molecule seems to be either completely untouched or completely disintegrated to soluble sugar. The reaction thus follows a pseudo first-order kinetics. Under some unusual circumstances, however, the reaction can be shifted to a bimodal type of degradation. Portions of the weak bonds in cellulose may, under these conditions, be preferentially cut to give some intermediate chains that reflect the corresponding crystallite structure. The situation can be further compounded by the possible multiple components of the cellulosic substrate as well as of the cellulase. More basic research is needed before a complete picture of the enzymatic hydrolysis of cellulose can be established.

Over the years, the production of cellulolytic enzymes has been studied in great detail but with limited progress. Most of this research has focused on the screening of possible mutants to help develop high cellulase producing microorganisms [80]. Understanding of enzymatic kinetics and the function and structure of cellulases is of special importance for the ultimate improvement of the enzyme activity. When this has been accomplished, researchers should be able to take advantage of the recent advances in recombinant DNA technology to carry out the cloning and amplification of cellulase genes. This would bypass the limitation of cellulase synthesis regulations by cellulolytic microorganisms and open the door for making the super-cellulase which would eventually facilitate the use of biomass as an energy source.

6 Nomenclature

β	β-bonds
a, b, c, θ	unit cell dimensional parameters
DP	degree of polymerization

Å	Ångström
LODP	leveling-off degree of polymerization
β_L-	looped β-bond
S1, S2, S3	three sublayers of the secondary wall
m	meter
g	gram
h	hour
°C	degree Celsius
C_x	endo-glucanase
C_1	exo-glucanase
C_b	cellobiase
E	enzyme
S	substrate
P	product
t	time
E_0	total enzyme concentration
k_1, k_2, k_{-1}	reaction rate constants
v	reaction rate
V	maximum velocity
S_0	initial substrate concentration
X_m	absorption parameter of enzyme
K_M	Michaelis-Menten constant
K, K′	reaction rate constants
A, B, D	constants
K_s	dissociation constant for ES complex
K_i, K_c	inhibition constants for enzyme and complex respectively
k_i	rate constant for ith component
S_i	substrate concentration of ith component
C	molecular cellulose
k	rate constant
C_i	molecular cellulose of component i
P_i	product produced from component i
α	fraction of cellulose molecules undergo consecutive reaction
I	intermediate chain length cellulose
k′, k″	rate constants
HC	hydrocellulose
α_i	fraction of cellulose component i undergo consecutive reaction
k_i', k_i''	rate constants of component i
$(HC)_i$	hydrocellulose from component i

7 References

1. Saeman, J. F.: Ind. Engr. Chem., Analyst. Ed. *37* (1), 43 (1945)
2. Meyer, K. H., Mark, H.: Ber. *61*, 593 (1928)
3. Meyer, K. H., Mark, H.: Z. Physik. Chem. *B2*, 115 (1929)
4. Meyer, K. H., Misch, L.: Helv. Chim. Acta *20*, 232 (1937)
5. Meyer, K. H.: Ber. *70*, 266 (1937)

6. Ellis, K. C., Warwicker, J. O.: I. Polymer Sci. *56*, 339 (1962)
7. Sarko, A., Muggli, R.: Macromolecules *7*, 486 (1974)
8. Gardner, K. H., Blackwell, J.: J. Biopolymers *13* (10), 1975 (1974)
9. Atalla, R. H., Nagel, S. C.: Science *185*, 522 (1974)
10. Kolpak, F. J., Blackwell, J.: Macromolecules *9*, 273 (1976)
11. Stipamovic, A. J., Sarko, A.: Macromolecules *9*, 851 (1976)
12. Chanzy, H., Dube, M., Marchessault, R. H.: Tappi, *61* (7), 81 (1978)
13. French, A. D.: Carbohydrate Research *61*, 67 (1978)
14. Buleon, A., Chanzy, H.: J. Polymer Sci., Polymer Physics *16*, 833 (1978)
15. Blackwell, J., Kolpak, F. J., Gardner, K. H.: Tappi *61* (1), 71 (1978)
16. Ellefsen, O., Tonnesen, B. A.: Cellulose and Cellulose Derivatives, Bikales and Segal (eds.), *IV*, p. 151. New York: John Wiley 1971
17. Manley, R. St. J.: Pulp Paper Report (PPRIC) *37*, 18 (1970)
18. Meyer, K. H.: Natural and Synthetic Polymers p. 312, Berlin: John Wiley 1950
19. Ingram, I.: Electron Diffraction and Microscopy of Cotton Cellulose, Paper presented at First Intern. Cotton Conf., Paris, April 1968
20. Scallan, A.: Textile Research J. *41*, 647 (1971)
21. Chang, M.: J. Polymer Sci. *C-36*, 343 (1971)
22. Chang, M., Tsao, G. T.: The Effect of Structure on Hydrolysis of Cellulose, Paper presented at 72nd AIChE Annual Meeting, San Francisco 1979
23. Sakaki, T., et al.: Biotech. Bioeng. *21*, 1031 (1979)
24. Tsao, G. T.: J. Chinese Chem. Eng. *9*, 1 (1978)
25. Addren, R. K.: Applied Polymer Symp. *28*, 205 (1975)
26. Cowling, E. B., Brown, W.: Adv. in Chem. Ser. *95*, 152 (1969)
27. Stone, J. E., Scallan, A. M., Donefer, E., Ahlgren, E.: Cellulases and Their Applications, Adv. in Chem. Ser. *95*, 219 (1969)
28. Millett, M. A., Effland, M. I., Caulfield, D. F.: Adv. in Chem. Ser. *181*, 71 (1979)
29. Earland, C., Raven, D. J.: J. Textile Inst. *2*, 69 (1977)
30. Cowling, E. B., Kirk, T. K.: Biotech. Bioeng. Symp. *6*, 95 (1976)
31. McIntosh, D. C.: Tappi *50*, 482 (1967)
32. Leopold, B., Monlik, S. K. R.: Tappi *51*, 334 (1968)
33. Kelsey, R. G., Shafizadeh, F.: Biotech. Bioeng. *22*, 1025 (1980)
34. Chou, T.: Ph. D. Dissertation, Purdue University, May 1980
35. Chang, M., Tsao, G. T.: Colloque Cellulolyse Microbienne, p. 27 CNRS Marseilles, France, (1980)
36. Jurasek, L.: Developments in Industrial Microbiology *20*, 177 (1979)
37. MacDonald, R. G.: Pulp Paper Manufacture Vol. 2, p. 453, New York: McGraw-Hill 1969
38. Seaman, J. E., Millett, M. A., Lawton, E. J.: Ind. Eng. Chem. *44*, 2848 (1952)
39. Sarkov, V. I., Dubush, O. Q.: Zh. Prikl Khim. *29* (9), 1411 (1956)
40. Sarkov, V. I., Levanova, V. P.: Gidroliz, Lesokhim. Prom. *13* (1), 5 (1960)
41. Welseth, G. S.: Tappi *35* (5), 233 (1952)
42. Ladisch, M. R., Ladisch, C. R., Tsao, G. T.: Science *201*, 743 (1978)
43. Hsu, T. A., Ladisch, M. R., and Tsao, G. T.: Chemtech *5*, 315 (1980)
44. Flickinger, M., Gong, C. S., Chen, L. F.: Conversion of Hemicellulose Carbohydrates, in: Adv. Biochem. Eng., Fiechter, A. (ed.), this volume.
45. Dunning, J. W., Lathrop, E. C.: Ind. Eng. Chem. *37* (1), 24 (1945)
46. Etheridge, D. E.: Can. J. Bot. *35*, 615 (1957)
47. Millett, M. A., Moore, W. E., Sàeman, J. F.: Ind. Eng. Chem. *46*, 1493 (1954)
48. Bertrand, D., Buleon, A.: Colloque Cellulolyse Microbienne, p. 127, CNRS, Marseille France, 1980
49. Cuggolz, J. R., Kohler, G. O., Klopfenstein, T.: J. Animal Sci. *33* (151), 167 (1971)
50. Mandel, M., et al.: Biotech. Bioeng. *14*, 1471 (1974)
51. Baker, A. J., Millet, M. A., Satter, L. D.: Cellulose Technol. Research, ACS Symp. Series *10*, p. 6, Washington, DC (1975)
52. Grethlein, H. E.: J. Appl. Chem. Biotech. *28* (4), 296 (1978)
53. Humphrey, A. E.: Adv. in Chem. Series *181*, 26 (1979)
54. Nisizawa, K. J.: J. Ferment. Technol. *51* (4), 267 (1973)

55. Wood, T. M.: Biotech. Bioeng. Symp. *5*, 111 (1975)
56. Ghose, T. K.: Adv. Biochem. Eng. *7*, 39 (1977)
57. Beriksson, K. E.: Biotech. Bioeng. *20*, 317 (1978)
58. Wood, T. M., McCrae, S. I.: Adv. in Chem. Series *181*, 181 (1979)
59. Hallowell, G., Giffin, M.: Biochem. J. *135*, 587 (1973)
60. Gong, C. S., Ladisch, M. R., Tsao, G. T.: Adv. in Chem. Series *181*, 261 (1979)
61. Mandel, M., Reese, E. T.: Develop. Ind. Microbial. *5*, 5 (1964)
62. Howell, J. A., Stuck, J. E.: Biotech. Bioeng. *17*, 873 (1975)
63. Huang, A. A.: Biotech. Bioeng. *17*, 1421 (1975)
64. Hsu, T. A., Tsao, G. T.: Biotech., Bioeng. *21*, 2235 (1979)
65. Ghose, T. K., Das, K.: Adv. in Biochem. Eng., Vol. *1*, p. 55, Berlin: Springer 1971
66. Van Dyke, B.: Ph. D. Dissertation, Mass. Inst. Technol., Sept. 1972
67. Brandt, D., Hontz, L., Mandel, M.: AIChE Symp. Series *69*, 127 (1973)
68. Reese, E. T., Segal, L., Tripp, V. W.: Txtile Research J. *27*, 629 (1957)
69. Reese, E. T.: Marine Boring and Fouling Organisms, Ray O. L. (ed.) p. 265, Seattle: University of Washington Press 1959
70. Amemura, A., Terui, G.: J. Fermen. Technol. *43*, 275 (1965)
71. Nisizawa, T., Suzuki, J., Nisizawa, K.: J. Ferment. Technol. *44*, 659 (1966)
72. Rinaudo, M., Bornoud, F., Merle, J. P.: J. Polymer Sci. *C-28*, 197 (1969)
73. Oginawa, Y., Arai, A.: J. Japan Technical Assn. Pulp Paper Ind. *21*, 209 (1967)
74. Philip, B., et al.: Adv. in Chem. Series *181*, 128 (1978)
75. Norkrans, B.: Physicol. Plantarum *3*, 75 (1950)
76. Cowling, E. B.: U.S.D.A., Technical Bulletin, No. 1258, p. 48 (1961)
77. Chang, M., Tsao, G. T.: Mechanism of Enzymatic Hydrolysis of Cellulose, paper presented at the 161th ACS Meeting, Atlanta 1981
78. Selby, K.: Adv. in Chem. Series *95*, 149 (1969)
79. Segal, L.: Cellulose and Cellulose Derivatives Part V, p. 733, Bikalis and Segal (eds.), New York: John Wiley 1971
80. Gallo, B. J., et al.: Biotech. Bioeng. Symp. *8*, 89 (1979)

Alcohol Production and Recovery

B. Maiorella, Ch. R. Wilke, H. W. Blanch
Lawrence Berkeley Laboratory and Department of Chemical Engineering
University of California Berkeley, California 94720

Alcohol production is receiving increasing attention as a potential transportation fuel. The main feedstocks for the fermentative production of ethanol by yeast are sugar, starch and cellulosic feedstocks. Conventional ethanol production technology by fermentation is reviewed, and the conventional distillation of ethanol from the broth is described. New advances in feedstock preparations including starch and cellulosic raw materials are reviewed and their impact on their fermentation is assessed. Continuous fermentation and continuous fermentation with cell recycle provide opportunities for more efficient alcohol production. Alternatives to energy intensive conventional distillation of alcohol include two-pressure distillation and various novel techniques such as dehydration and solvent extraction. A review of alcohol production economics is included.

1 Introduction

1.1 Conventional Ethanol Technology

The total world industrial ethanol production is approximately 3,2 billion liters (800 million gallons) per year [1]. Three quarters of this production uses the same techniques developed prior to the 1940s; slow batch process, followed by efficient multistep distillation [2]. The conventional practice is well illustrated in the United States Department of Agriculture Peoria alcohol pilot plant (built in 1940) to convert 5,5 t (200 bushels) per day of corn or grain to 2000 l (500 gallons) per day of anhydrous ethanol [3].

1.2 Conventional Process

The overall conversion process for corn is given schematically in Fig. 1. Corn or grain is first received and milled, then pulped by pressure cooking (horizontal cookers, Fig. 2). Malt is then added to the pulped mash to hydrolyze the plant starches and produce sugar for ethanol production. A high degree of conversion to sugar is essential for good alcohol economics as the corn raw material cost contributes up to 70% of the final alcohol cost. Hydrolyzed mash is diluted with water to give a 15 wt% sugar solution which is then gravity fed to the fermentors below (Fig. 2 ground level).

Batch process is a slow reaction. The vessel is first cleaned and prepared. The sugar solution, supplemented with yeast nutrients, is added to the vessel and the vessel is inoculated with a rapidly growing culture of yeast from a seed tank. The course of a process using a typical yeast strain is plotted in Fig. 3. The inoculated yeast continue to multiply, simultaneously producing alcohol. The rate of alcohol production is initially quite low, but as the number of yeast cells increases, the overall rate increases. A maximum in ethanol productivity is reached after 20 hours. Ethanol production then continues at a decreasing rate until at 28 h, 94% of the sugar is utilized, and a final ethanol content of 46 g l^{-1} ethanol is achieved. The reaction mixture is then emptied into a holding tank. The vessel is cleaned and prepared for another process cycle. Several fermentors are operated at staggered intervals to provide a continuous feed to the distillation system.

The overall productivity for the batch process is typically only 1.8 to 2.5 g of ethanol produced per l of fermentor volume per h [3,4] and eighteen 100,000 l batch fermentor are required for an industrial scale (100 million l per year) alcohol fuel plant.

1.3 Distillation

Distillation of the dilute alcohol beer is carried out in several stages. First, weak alcohol vapors are steam stripped from the beer in the beer column (leftmost column, Fig. 4). The dilute alcohol vapors are then fed to the aldehyde column (middle of

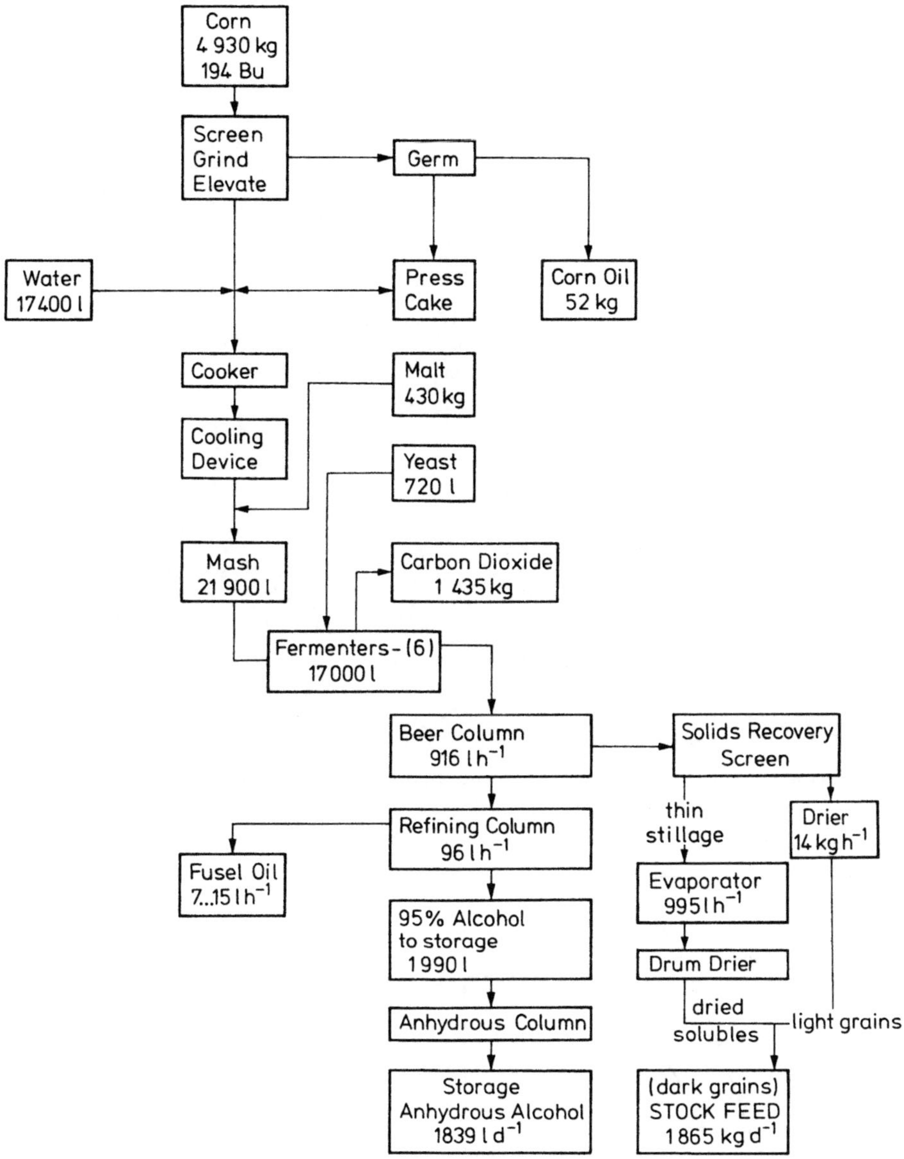

Fig. 1. Peoria corn to alcohol pilot plant; process flow diagram

Fig. 4) where aldehyde (a by-product) is stripped off and removed through the top condenser, and fusel oils are removed as a side draw from the lower plates. The refined alcohol is then further concentrated in a small rectifying column (hidden behind the middle column) to produce near azeotropic (95 wt%) alcohol. To produce pure alcohol, the 95 wt% product is further distilled (in the right hand column) using ether at 690 kilopascals (100 psi) as an azeotroping agent. The final

Fig. 2. Peoria pilot plant; fermentation section (by courtesy of USDA)

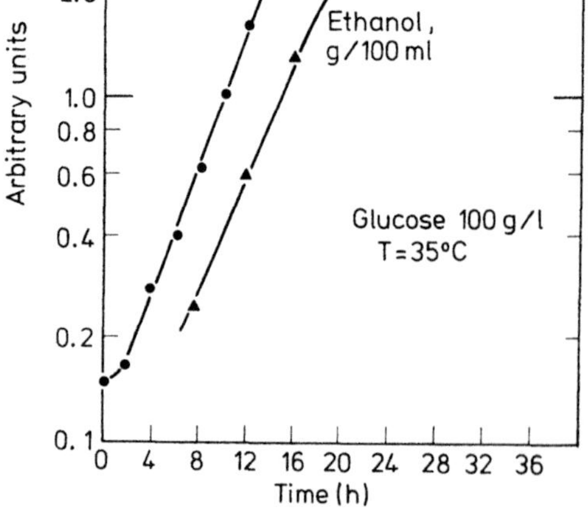

Fig. 3. Typical batch ethanol production data

Fig. 4. Peoria pilot plant; distillation section (by courtesy of USDA)

pure alcohol is taken from the bottom of the tower. This overall distillation process is quite efficient, using approximately 2.5 kg of steam to produce each liter of pure alcohol product.

1.4 By-Product Recovery

By-product recovery is important to favorable alcohol economics. The stillage from the beer column contains corn residue and dead yeast. This material is screened and the solids are dewatered in a screen press to produce high protein distillers light grains. The liquid (high in non-degradable sugars and other carbohydrates) is concentrated in the triple effect evaporator (Fig. 4, right side) to produce a heavy syrup-distillers solubles. Both distillers light grains and distillers solubles are valuable as cattle feed.

1.5 Conclusions

Several important conclusions can be drawn from this overview of the Peoria plant and these establish the emphasis of this paper.

Raw material costs make up the major portion of final alcohol cost. The selection of a raw material is central to the economics of any new alcohol plant. A high percentage conversion of any raw material to sugars and thence to alcohol is crucial.

Conventional technology is highly inefficient. Designs for full scale (35 to 200 million l per year) alcohol plants are available in the literature [2,5,6,7]. Owing to the hugh volume requirements of the reactors, finance and depreciation costs are very high, contributing about 2.6 cents per l (10 cents per gallon) to the final cost of alcohol for the fermentation section of the plant alone. The continual start up/shut down nature of the batch process makes it difficult to automate and high labor costs also result. The need for new processes is apparent. Fortunately, great progress has been made with as much new research conducted over the past fifteen years as in the previous fifty [8].

Process by-products are quite valuable. Yeast sold as a feed supplement is valued at 8–20 cents per kg and can reduce final alcohol cost by as much as 5.8 cents per l (22 cents per gallon). Consideration of all by-products for potential value is important.

Finally, distillation has recently been considered to be very energy inefficient-utilizing as much energy as is produced in the alcohol product. This is shown to be clearly unfounded in the Peoria pilot plant where the energy consumed for distillation is less than 30% of the product energy. Several distillation alternatives with different special advantages do exist and should be compared.

2 Feed Stocks

2.1 Feedstock Alternatives

Sugar for alcohol production (sucrose, glucose or fructose) may be derived from any of three major classes of raw materials: sugar containing materials, starch materials, and cellulosics. Sugar containing materials require the least costly preparation, as sugars are already available in degradable form. This class of materials is, however, the most expensive to obtain. Starchy materials are cheaper, but require processing to convert the starch to fermentable sugar. Cellulosic materials are available as cheap waste residues, but require extensive and costly processing.

2.2 Sugar Containing Feedstocks

Sugar cane juice and black strap molasses are the most important of the sugar containing raw materials. These are used almost exclusively in Brazil (where 2.5 million m^3 of alcohol per year are produced for motor fuel blending) [9] and were also the major sugar sources for U.S. industrial alcohol production during World War II [10].

Sucrose ($C_{12}H_{22}O_{11}$) occurs in free form in sugar cane. Sugar cane juice is prepared by grinding and extracting the sugar cane followed by a clarification process whereby inorganics are precipitated with dilute sulfuric acid followed by lime. Clarified cane juice may be used directly for reaction. Black strap molasses is a by-product of table sugar production. In the production of table sugar, sucrose from sugar cane is crystallized by multi-effect evaporation. A noncrystallizable residue (mainly glucose and fructose) results. This residue (50–60% sugars in water) is black strap molasses. Molasses, concentrated in sugar, is easily stored. Dilution to 10–20 wt/vol.% before reaction is typical.

Sugar beets and sweet sorgum have potential as sugar containing raw materials for ethanol production. Fruit cannery waste is another possible source.

2.3 Starchy Feed Materials

Starchy raw materials are either cereals (corn, wheat, oats, rice, etc.) or starchy root plants (sweet potatoes and cassava). The cereal grains offer the advantage of easy year-round storage and, corn (containing 50–60% starch) is the most prevalent starchy raw material used in the U.S.

Cereal starch is converted to sugar by enzymatic hydrolysis. The grain is first milled and pressure cooked to solubilize the starch. After cooling, saccharifying enzymes are added. The enzyme may be derived from molds (the amylo process) or from germinating barley (malt processes). The batch hydrolysis process completes the conversion to maltose and dextrin sugars in one to two hours. Continuous cooking and flash conversion process has been developed, however, which can complete the starch to sugar conversion in a matter of minutes [10].

Potatoes and cassava require hydrolysis processes similar to the grains. These crops offer a great advantage in giving a far higher yield of starch per ha of crop land, but the problem of storage to allow year round alcohol production has not been overcome.

2.4 Cellulosic Raw Materials

Cellulosic raw materials are available as wastes (waste newsprint, etc.), as farm residue (corn stalks, sugar cane bagasse, wheat and rice straw, saw mill waste) or as harvestable wood. Cellulose is, however, difficult to convert efficiently to its degradable monomer sugar units. Acid [11,12,21] and enzymatic [13,14,15] hydrolysis processes are under intensive laboratory development, but conversion yields are generally limited to less than 60% [15] and these processes are not yet economical. Substantial improvement may be imminent.

2.5 Feedstock Comparison

For conventional feedstocks, raw material price, conversion efficiencies and by-product sales credits (from stillage for cattle feed) are well known. Net raw materials costs per l of alcohol product can easily be computed. For cellulosic feeds, conversion efficiencies are not well known. Very little is known of potential by-product values,

Table 1. Alternate sugar sources, substrate costs

Item	Units	Corn	Grain sorghum	Wheat	Cull potatoes	Sugar beets	Starch	Molasses	Corn stover	Douglas fir
Raw material price	dollars	2.50/bu**)	2.30/bu	3.35/bu	20/ton	26/ton	0.16/kg	0.09/l	$ 50/ton	$ 15/ton
Conversion	l ethanol	9.9/bu	9.9/bu	9.9/bu	109/ton	76.9/ton	0.5/kg	4/l	216/ton*	273/ton*
Unit raw material cost	$/l ethanol	0.25	0.23	0.34	0.18	0.34	0.32	0.26	0.23*	0.06*
By-product price	dollars	110/ton	110/ton	110/ton	6/ton	93/ton	0	15/ton		
Conversion	kg by-product	7.64/bu	7.64/bu	9.41/bu	135/ton	115/ton	.05/kg	1.87/l		
Unit by-product credit	$/l ethanol	0.095	0.095	0.12	0.008	0.16	0	0.032		
Net raw material cost	$/l ethanol	0.155	0.135	0.220	0.172	0.180	0.320	0.228		

* Maximum yield assuming 100% conversion of cellulose to glucose — Typical yields are 1/3 of this

** 1 bushel = 35 l grain

and even the raw material costs themselves are disputed. Conversion costs for all but the industrialized processes of WW II are unestablished. Available substrate cost information is summarized in Table 1.

3 Ethanol Formation

Practice in most plants is essentially unchanged from the low productivity batch processes of WW II. Much new research shows, however, that substantial improvements are possible.

Selected high productivity yeast strains are used almost exclusively for ethanol production. Yeast produce ethanol with very high selectivity (only traces of by-products) and are extremely hardy. Other organisms have been studied for use in ethanol formation. *Clostridium thermosaccharolyticum* and other thermophilic bacteria have been considered since high temperature reaction would allow simple continuous stripping of ethanol product from the active reaction mixture, thus eliminating product inhibition effects [16]. Organisms so far studied, however, produce excessive quantities of undesirable by-products. An important possibility for the future is the development of an organism especially tuned to rapid ethanol production. Genetic modification of the simple bacterium *Zymomonas mobilis* to give very high product concentration without inhibition is currently being examined [17].

3.1 The Theory of Metabolism of Yeasts

An understanding of the basic metabolic reactions and their regulation is helpful in the evaluation of alternative processes.

3.1.1 Yeast Metabolic Pathways

A schematic representation of the anaerobic and aerobic yeast metabolic pathways is presented in Fig. 5. Under anaerobic conditions, glucose is converted to ethanol and carbon dioxide by glycolysis. The overall reaction to liberate energy for biosynthesis (Eq. (1)) produce 2 moles of ethanol and carbon dioxide for every mole of glucose consumed.

Theoretical:

$$C_6H_{12}O_6 \rightarrow 2\,C_2H_5OH + 2\,CO_2 + \text{Energy (stored as ATP)}\,. \qquad (1)$$

On a weight basis, every gram of glucose can theoretically yield 0.51 g of ethanol. In practice though, actual ethanol yield are about 90% of the theoretical — a portion of the glucose carbon source being used for synthesis of new cell mass (Eq. (2)).

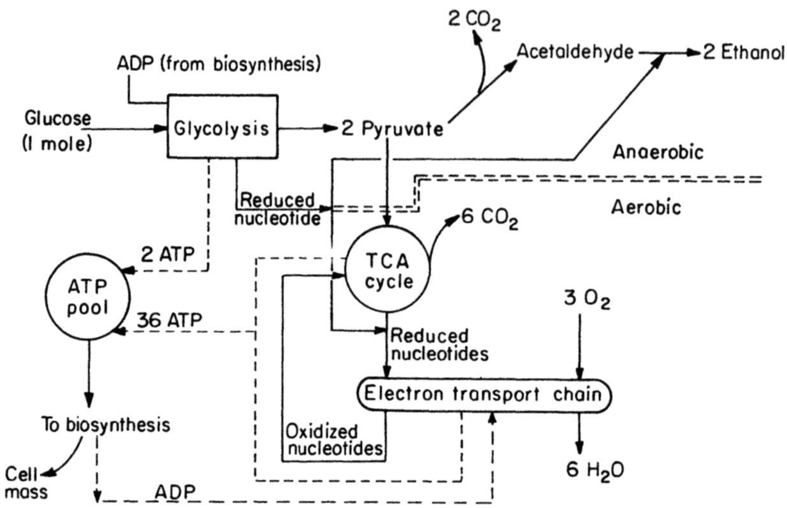

Fig. 5. Simplified chart of anaerobic and aerobic catabolism of *Saccharomyces cerevisiae*

Observed:

$$\text{lg } (C_1H_{12}O_6) \rightarrow 0.460 \text{ g } (C_1H_5OH) + 0.440 \text{ g } (CO_2) + 0.10 \text{ g (new cells)} . \tag{2}$$

Under aerobic conditions, glucose is converted completely to carbon dioxide and new cell mass, with no ethanol being formed.

3.1.2 Effect of Glucose Concentration

Glucose is the primary reactant in the yeast metabolism and at very low glucose concentrations (less than 10 mg l^{-1}) the rate of glucose consumption increases roughly linearly with glucose concentration [18]. At high concentration (above 150 g l^{-1}) glucose inhibits enzymes in both the repressed and oxidative pathays [19]. At intermediate glucose concentration (3 to 100 g l^{-1}) catabolite repression of the oxidative pathways takes place permitting production of ethanol even in the presence of large amounts of oxygen [20,21,22].

Under anerobic growth conditions, caused either by catabolite repression or by oxygen restrictions, and at low to moderate glucose levels, the rate of ethanol production is given roughly by a Monod type relationship

$$V = V_{max}S/(K_s + S) ,$$

where:

- V is the ethanol production rate in g ethanol/g cells-hr .
- S is glucose substrate concentration (g l^{-1}) .
- K_s is a constant factor, having a very low value on the order of 0.35 g l^{-1} ,

so that under anaerobic conditions the rate of ethanol productivity per cell is essentially at its maximum value for sugar concentrations considerably above 3.5 g l^{-1}, and less than 150 g l^{-1} (where inhibition becomes important).

3.1.3 Effect of Oxygen

It is important to avoid aerobic metabolism which utilizes glucose feed but produces no ethanol. At high glucose concentration, catabolite repression assures ethanol formation. At low sugar concentration, the oxygen concentration must be restricted. Oxygen ordinarily should not be totally eliminated, however, as oxygen is a necessary building block for the biosynthesis of polyunsaturated fats and lipids required in mitochondria and plasma membrane [23,24,25,26]. Trace amounts of oxygen (0.5 to 0.10 mm Hg oxygen tension) in the reaction mixture are adequate [27,20,28,24,30] and do not promote aerobic metabolism. As oxygen is continually consumed by the yeast during growth, it is normally necessary to sparge the fermentor with air to maintain even these trace oxygen levels.

3.1.4 The Effect of Ethanol

Ethanol, the main product, is itself a yeast toxin. The tolerance of various yeast to ethanol depends considerably on the strain chosen, but ethanol production and cell growth are generally brought to a complete halt at ethanol levels greater than 110 g l^{-1} [27,31,32]. For most yeast, the effect of ethanol inhibition is negligible at low alcohol concentrations (less than 20 g l^{-1}) but increases rapidly at high concentrations. (Fig. 6a, b) [32,33,34,35].

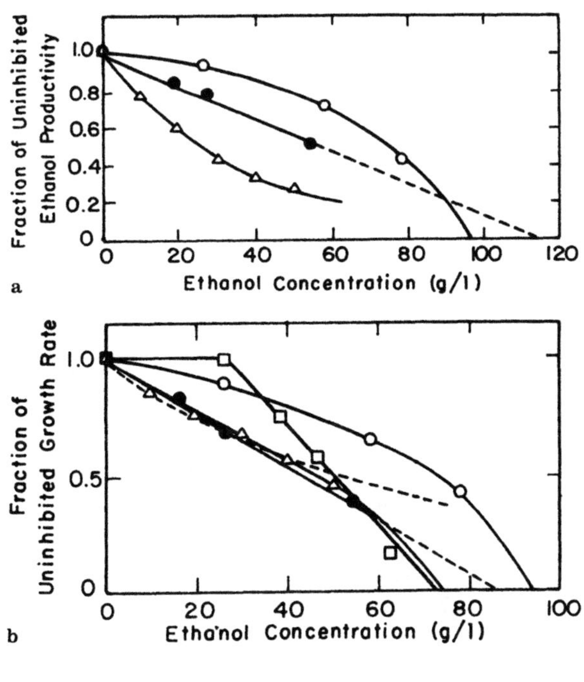

Fig. 6a. Ethanol end product inhibition (○ Bazua and Wilke, △ Aiba, ● Ghose and Tyagi). **b** Ethanol inhibition of cell growth (□ Holzberg, ○ Bazua and Wilke, — Aiba, △ Aiba and Shoda, ● Ghose and Tyagi)

3.1.5 Additional Nutrient Requirements

In addition to providing a sugar source for conversion to ethanol, media must also provide the nutrients to support cell growth. The major requirements are for the major components of the cell — carbon, oxygen, nitrogen, and hydrogen. Lesser quantities are required for the minor cell components — phosphorous, sulfur, potassium, and magnesium. Even less are required of the trace minerals (given in Table 2 [36]) and the organic growth factors, which comprise amino acids, purines and pyrimidines and vitamins. The most important growth factors for yeast are the vitamins, biotin, pantothenic acid, inositol, thiamine, nicotinic acid, and folic acid [37,38]. A typical synthetic medium for ethanol production by yeast is given in Table 3 [172] and shows the approximate proportions required for balanced growth. A balanced medium is especially important for closed or partially closed system in which excess elements can build up to toxic levels..However, more exact medium requirements have to be determined experimentally for the individual organisms. The extent of variation in media can be seen in the wide range of concentrations given in Oura's [39] summary of twelve of the best known yeast culture media.

In industrial scale, however, the carbon, hydrogen, and oxygen are normally provided by a complex carbohydrate source, such as cane or beet molasses. Aside from the nitrogen present in the amino acids, which normally accompany the carbon source, nitrogen is usually supplied as ammonia or ammonium salts, particularly ammonium sulphate. Urea is also used for economic reasons, but it is less readily assimilated unless biotin is also added. Phosphorous is provided as phosphoric acid or ammonium or potassium phosphate, which also provides the potassium. The other media components are normally present with the carbon sources, although the magnesium, sulphates, chloride, biotin, and thiamin may need supplementation. The composition for molasses is given in Table 4 [168] from which the supplementary nutrient requirements for it or any other carbon source of known composition can be calculated from the synthetic requirements given in Table 3.

Table 2. General media trace element requirements

Most likely required:	Mn, Co, Cu, Zn
Maybe required:	B, Al, Si, Cl, V, Cr, Ni, As, So, Mo, Sn, I
Least likely:	Be, F, Sc, Ti, Ga, Ge, Bi, Zr, W

Table 3. Composition of a typical synthetic medium

Glucose	100.00 $g\,l^{-1}$	KI	1.0 $mg\,l^{-1}$
$(NH_4)_2SO_4$	5.19 $g\,l^{-1}$	$FeSO_4 \cdot 7\,H_2O$	2.0 $mg\,l^{-1}$
KH_2PO_4	1.53 $g\,l^{-1}$	$Al_2(SO_4)_3$	3.0 $mg\,l^{-1}$
$MgSO_4 \cdot 7\,H_2O$	0.55 $g\,l^{-1}$	Biotin	0.125 $mg\,l^{-1}$
$CaCL_2 \cdot 2\,H_2O$	0.13 $g\,l^{-1}$	Pantothenate	6.25 $mg\,l^{-1}$
H_3BO_3	10.0 $mg\,l^{-1}$	Inositol	125.0 $mg\,l^{-1}$
$CoSO_4 \cdot 7\,H_2O$	1.0 $mg\,l^{-1}$	Thiamine	5.0 $mg\,l^{-1}$
$CuSO_4 \cdot 5\,H_2O$	4.0 $mg\,l^{-1}$	Pyridoxine	6.25 $mg\,l^{-1}$
$ZnSO_4 \cdot 7\,H_2O$	10.0 $mg\,l^{-1}$	p-Aminobenzoic Acid	1.0 $mg\,l^{-1}$
$MnSO_4 \cdot H_2O$	3.0 $mg\,l^{-1}$	Nicotinic acid	5.0 $mg\,l^{-1}$

Table 4. Composition of molasses

Element	Molasses (mg per 100 g water free basis)	Element	Molasses (mg per 100 g water free basis)
C	$3.9–4.1 \times 10^4$	Cu	0.1–6
H	$6.3–6.8 \times 10^3$	Mn	1–4
O	$4.7–5.0 \times 10^4$	Co	0.04–0.1
N	$0.1–2.8 \times 10^3$	Mo	0.009–0.026
K	$0.8–5.2 \times 10^3$	Cl	1300
P	10–900	I	—
S	150–200	Pb	0.6
Mg	7–750	As	—
Ca	18–1200	Si	28
Na	82–1400	Sr	5
Zn	0.6–130	B	0.2–0.4
Fe	10–21		

Vitamin content

Vitamin	Molasses $\mu g\ g^{-1}$ liquid molasses
Thiamine	0.8
Riboflavin	—
Nicotinic acid	15
Panthothenic acid	20
Pyridoxin	—
Biotin	1.5
Inositol	2000

3.1.6 Total Versus Specific Productivity

Thus far, we have discussed specific productivity, the rate of ethanol production per yeast cell. Specific productivity is a function of glucose concentration, oxygen concentration and ethanol concentration (as well as of many independently controllable variables such as temperature and pH). An industrial process must maximize total productivity, which is the product of specific productivity and yeast cell concentration.

Typical cell concentrations at the end of a batch are of the order of only 5–10 $g\ l^{-1}$. Continuous experiments have been conducted at levels up to 120 $g\ l^{-1}$, and the expected increases in total ethanol productivity have been observed [41,42]. At higher cell densities, approaching the maximum packing density of 200 $g\ l^{-1}$ [43], transport limitations [44] and possible cellular inhibition are anticipated [45,46].

3.2 High Rate Processes

3.2.1 Conditions for High Rate Turnover

With the basic yeast fermentation reaction and its regulation by ethanol, oxygen and glucose concentrations outlined, we can now propose those conditions which would lead to a high rate process.

The specific productivity must be high. Therefore, the oxygen level should be regulated to prevent aerobic growth, but to adequately meet yeast oxygen maintenance requirements. The ethanol level should be maintained low, so as not to inhibit further production. Glucose concentrations should be maintained at not less than 3.5 g l^{-1} so as not to inflict a feed limitation. Finally the concentration of yeast cells (the catalyst for the ethanol production reaction) should be maintained as high as possible without causing transport associated limitations.

3.2.2 Criteria for Evaluation of Potential Industrial Processes

Very few of the high rate processes under development have been advanced to the point of pilot plant testing. Detailed economic evaluations of the potential processes are therefore not available and another means of comparison is required. Some simple criteria for evaluating the potential processes for industrial applications are pesented in Table 5.

Table 5. Some criteria for comparing ethanol processes

Low operating cost:	*Low capital cost*:
1. Continuous process	1. High productivity (small reactor volume)
2. Simple operation	2. Mechanically simple
3. Low energy input	
4. Near complete sugar utilization	

Fully continuous processes offer many advantages [47,48,49]. 1) The process can be conducted at a single optimum condition (not varying with time); 2) Greater throughput of product and consistency of product quality are possible; 3) Total production is increased and downtime between cycles is eliminated; 4) A steady demand is placed on electricity, steam and other services, thus eliminating high peak load levels; 5) Automation is simplified based on real time sampling of the final product; 6) Intermediate product storage is not required for buffering to interface with continuous feed pasteurization and product concentration units; 7) The process is more easily controlled and adjusted to meet varying requirements. For these many reasons, continuous processes are to be favored for industrial applications.

Simple processes (with few steps) reduce maintenance requirements. Control is made simpler and these processes should be favored.

Energy and glucose feed costs contribute substantially to the cost of the final product. Processes which utilize little energy (for pumps, refrigeration, etc.) and which utilize the glucose feed as completely as possible (minimizing waste) are clearly advantageous.

The special advantages of high rate fermentation processes come in reduced capital and maintenance costs. High-rate processes allow the use of much smaller equipment. When the process is also mechanically simple, major capital cost savings are assured.

3.3 Alternative Processes

3.3.1 The Simple Continuous Stirred Tank Fermentor

The continuous stirred tank reactor (C.S.T.R.) is a simple point of departure from standard batch process.

A C.S.T.R. is depicted schematically in Fig. 7. Pasteurized feed is pumped continuously into an agitated vessel in which yeast are actively fermenting. The sugar is largely consumed and ethanol and new cell mass are produced. Beer containing ethanol, yeast cells, and residual sugar, flows continuously from an overflow port in the side of the vessel. Air is sparged through the vessel base to maintain the optimum oxygen tension. The composition of the beer in the reactor is everywhere uniform and is the same as the composition of the overflow stream [50)].

Specific ethanol productivity in the simple C.S.T.R. fermentor is ordinarily limited by ethanol inhibition [41)]. To avoid very high ethanol distillation costs, the ethanol overflow product concentration and hence the concentration in the fermentor must be maintained at a relatively high level. Cysewski has found an optimum of 5 w/v% to give minimum bioconversion plus distillation cost for one yeast strain [5)].

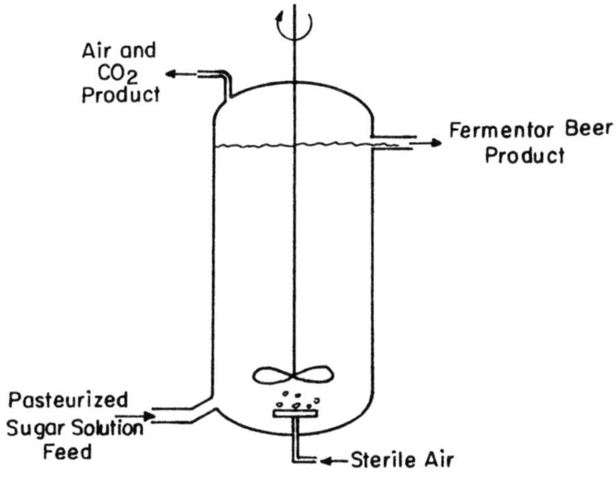

Fig. 7. Continuous stirred tank reactor (CSTR)

The total productivity of a C.S.T.R. system is also limited by the low cell density achieved in the fermentor. New cells are continuously born in the vessel, but cells are also continuously washed out. A steady state is achieved when the growth and washout rates are identical. A cell density of only 10–12 g l^{-1} is typical [41)]. The overall productivity for a simple C.S.T.R. using a high productivity yeast is approximately 6 g l^{-1} h^{-1} ethanol, three times the average batch productivity [5,171)].

Considered in terms of our evaluation criteria the simple C.S.T.R. process offers many advantages over conventional batch technology. The process is continuous, involves few steps, has no unusual energy requirements and can achieve virtually complete sugar utilization. Reduced operating costs are expected. The equipment is mechanically very simple and total required volumes are only 1/3 of those required for batch operation. Capital costs should therefore be reduced. These predictions are supported by a detailed economic analysis by Cysewski which shows a 53% reduction in operating costs and a 50% reduction in capital costs for the C.S.T.R. process as compared to conventional batch reactions [5].

3.3.2 Series C.S.T.R.s

Advantages have been demonstrated for C.S.T.R.s arranged in series both at laboratory [33,47,48] and plant scale [2,51]. Consider first, two C.S.T.R.s in series. Sugar solution is fed to the first and reaction takes place. The residence time is adjusted so that the sugar is only partially utilized in the first vessel. The ethanol concentration is thus less than for complete utilization. The overflow from the first vessel is fed to the second, where the reaction is completed producing the final high ethanol concentration beer product. Because the first fermentor is operated at reduced ethanol concentration, ethanol inhibition is reduced and the productivity of the first fermentor is quite high. The second, lower productivity fermentor, now must convert less sugar than if it was operated alone. The result is an overall increase in productivity compared to a single vessel. Ghose and Tyagi [33] have shown the productivity of a two-stage C.S.T.R. system to be up to 2.3 times that of a single C.S.T.R.

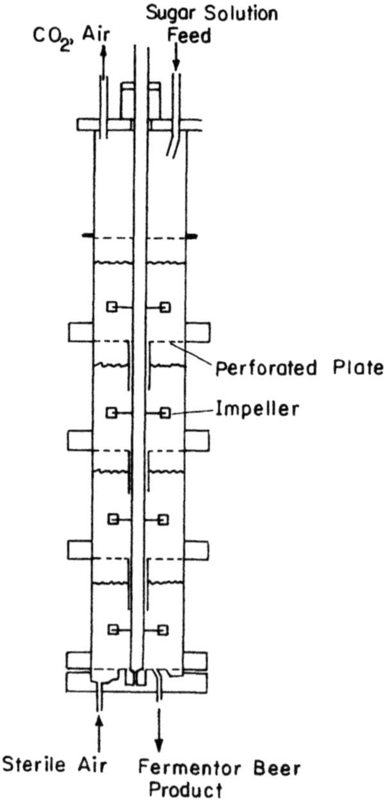

Fig. 8. Multistage perforated plate column fermentor

Systems with several vessels in series can also be considered. The concept of multiple series C.S.T.R.s has been applied in the multistage perforated plate column fermentor (Fig. 8) [52,53,54,55,56]. The reactor consists of a column divided into stages by perforated plates. A single shaft drives impellers at all stages. Each stage acts as an individual vessel. Feed added at the column head is partially converted in the first stage and it trickles down and is successively fermented through the lower stages.

Like the simple C.S.T.R. system, the simple two-stage series C.S.T.R. system appears advantageous based on the selection criteria. It is not clear though, whether the increase in productivity and the overall reduction in volume will be justified by the added complications of a multiple vessel system.

The mechanical complexity of the perforated plate column reactor is a major factor against its industrial application.

3.3.3 Continuous Cell Recycle Reactors

The use of continuous cell recycle reactors has been thoroughly investigated [5,57,32,58] and has been applied at large scale [59]. A cell recycle system is shown in Fig. 9. The system is identical to the simple C.S.T.R.s, except that a centrifuge is used to separate yeast from the product overflow and return this yeast to the vessel. With continued cell growth, and cell escape prevented, the cell concentration in the reactor becomes extremely high and total productivity is greatly increased. A small bleed of cells is required to maintain a viable culture [18,60], but cell densities as high as 83 g l^{-1} can be maintained [61,42]. Higher concentration sugar feeds can be fermented, and productivities of 30–40 g l^{-1} h^{-1} are possible [33,41,42,62].

This continuous process is also quite attractive. Some added complexity results from the need for a mechanical centrifuge which increases capital cost and requires considerable maintenance. Electrical energy costs are increased, and added supervision is required to monitor the centrifuge. These disadvantages have been shown to be more than offset by the great increase in productivity and reduction in equipment size [57].

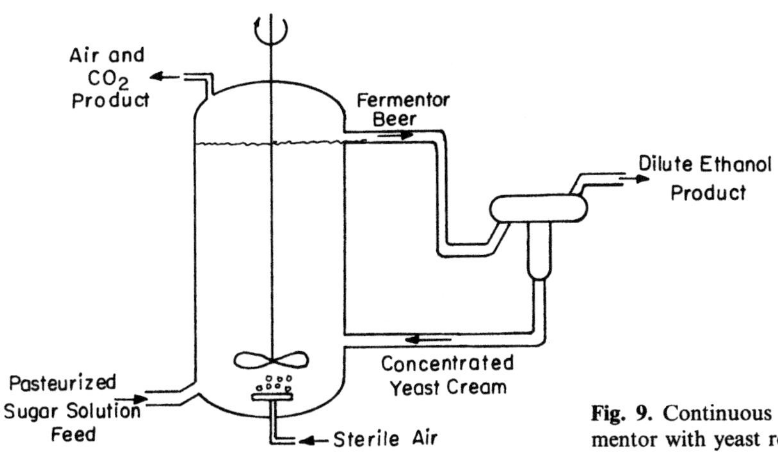

Fig. 9. Continuous stirred tank fermentor with yeast recycle

Several attempts have been made to develop simplified cell recycle systems which do not require mechanical centrifuges. Simple cell settling systems have been proposed wherein the cells are thermally shocked (to temporarily halt CO_2 evolution) and allowed to gravity settle. Very large settling vessels are required, however [63,64]. Whirlpool separators have been investigated (Fig. 10) [65]. Yeast cells are deposited in a central cone when the overflow is pumped tangentially into a vertical cylindrical vessel. Energy requirements for the whirlpool separator are low [66].

A very simple partial recycle fermentor has been developed and tested at pilot scale (Fig. 11) [67]. The overflow is taken from a vertical pipe rising through the vessel base. This pipe is jacketed by a baffled sleeve. The region between pipe and sleeve escapes agitation and yeast tend to separate from the beer rising to the overflow nozzle. The separation is far from complete, but some concentration of cells is achieved, and with essentially no added equipment.

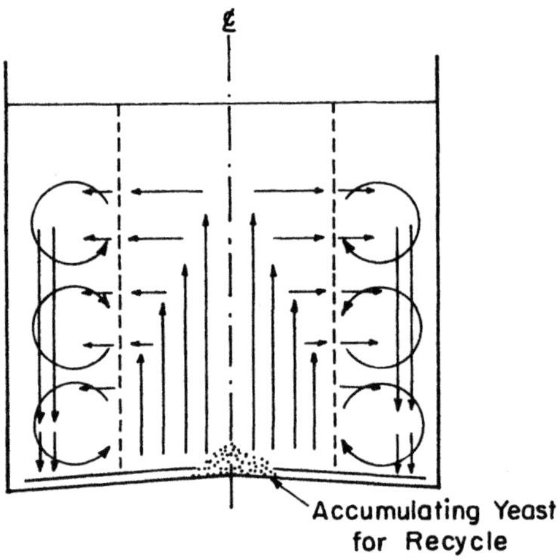

Fig. 10. Whirlpool yeast separator

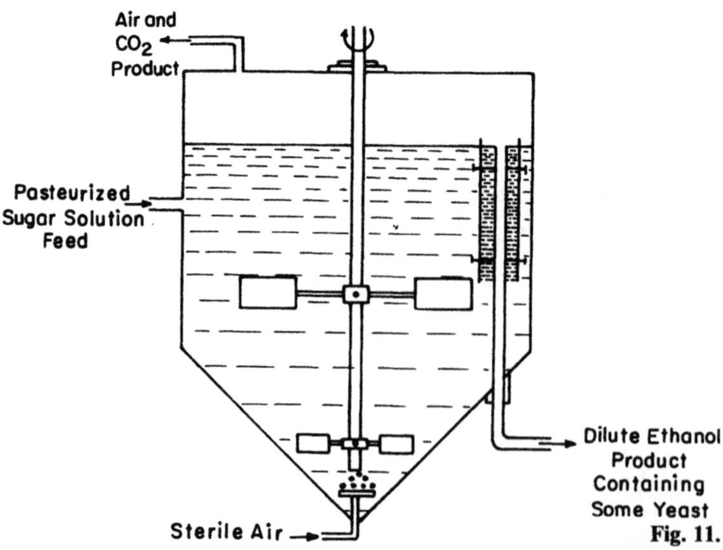

Fig. 11. Partial recycle reactor

3.3.4 Tower Fermentors

Tower fermentors of many designs have been tested. The APV tower system (Fig. 12) has been operated successfully at large scale [68]. The reactor consists of a vertical cylindrical tower with a conical bottom. The tower is topped by a large diameter settling zone fitted with baffles. The overall aspect ratio is from 7:1 to 10:1 with tower diameters from 0.9 to 2 meters [69, 70, 71, 72]. Sugar solution is pumped into the base of the tower, which contains a plug of flocculent yeast [68]. Reaction proceeds progressively as the beer rises, but yeast tends to settle back and be retained. The limitation on beer throughput is set by the requirement that yeast be effectively retained by settling against the upward flow. High cell densities of 50 to 80 g l^{-1} are achieved [68] without the requirements of an auxiliary mechanical separator. Productivities 32 to 80 times those for simple batch processes have been achieved with the APV tower system [2]. Caution must be used in comparing productivities for various process schemes. Different yeasts and sugar feed solutions have been used in testing the various systems. Productivities 80 times higher than for a simple batch have been achieved in the tower fermentor comparing slow growing brewers yeast in both cases. The productivity of 83 g l^{-1} h^{-1} ethanol achieved by Cysewski [41] with vacuum process is 40 times the batch productivity achieved with the same yeast — yet

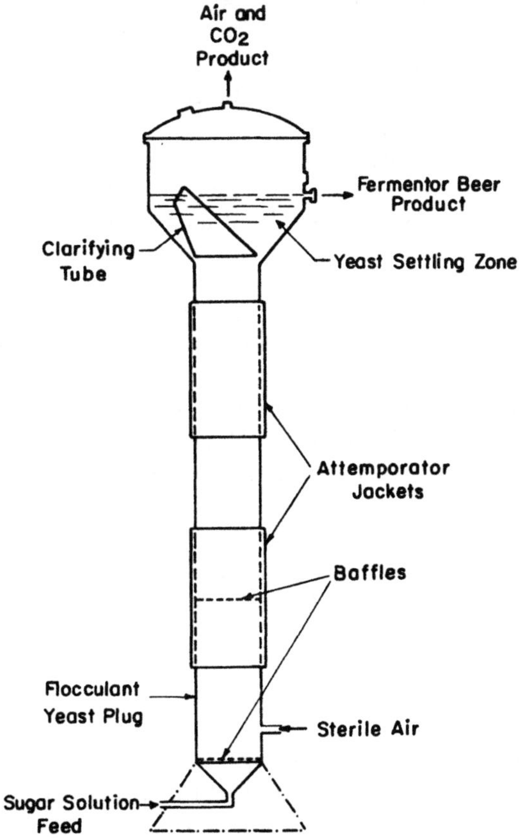

Fig. 12. APV tower fermentor

this is the highest absolute conversion rate reported. Some apparent difficulty is associated with providing the desired oxygen concentration in the tower fermentor as direct sparging at the base promotes turbulence and interferes with yeast settling.

A major drawback of the APV system is the long time required for initial start-up. Two to three weeks are required to build up the desired high cell density and achieve stable operation [2]. This is compensated by the very long — 12 months and greater — run times between shut-downs.

Based on the simple process selection criteria, the tower fermentor again appears to be an improvement over the processes discussed aerlier. The system is very simple. While added power is required for pumping the beer up through the tower, this is offset by the savings from the elimination of a agitator. The high productivity achieved in the APV system results in very small vessels. Capital and operating costs are predicted to be far lower than for conventional batch processes.

The slant tube fermentor has been proposed as a modification to the tower fermentor with a higher beer flow rate capacity [73]. A three cm diameter tube, 14 m long, is mounted at a 45 degree angle to horizontal. Sugar solution is pumped through the tube from the base. Reaction takes place progressively up the tube. Retention of yeast in the slant tube at high flow rates is possible as the yeast settling depth is very shallow — yeast need only settle to the tube lower wall. Once settled, yeast roll rapidly down the tube. As CO_2 is evolved it rises to the tube upper wall where it bubbles rapidly up and out of the vessel. A three phase flow results with CO_2 flow separated and therefore no longer interfering with yeast settling. In a laboratory study a productivity of 25 $g\,l^{-1}\,h^{-1}$ was acieved. This rate is 39 times greater than that for a simple batch with the same grape juice feed [73].

A bank of many slant fermentor tubes operated in parallel would be required to achieve industrial scale production. Laboratory tests have shown that a single pump feeding several parallel tubes from a single manifold could not mantain equal flow rates in them all [73], and a special flow distributor system will be required for an industrial plant.

While the slant tube fermentor may allow somewhat increased productivity over the simple APV tower design, the added complexity of a multiple tube system will probably offset this advantage.

Another approach to allowing increased throughput in tower fermentors has been to pack the vessel with standard distillation column type packing [74]. The packing material provides zones of quiescence where yeast cells can collect and very high cell densities can be maintained even at high flow rates. Continual cell growth can cause selective plugging and fluid by-passing so that the column must be frequently sparged with a rapid nitrogen flow to shake free and redistribute the cells [74]. This disrupts stable continuous operation and may mitigate the advantages of slightly increased productivity.

3.3.5 Dialysis Fermentors

Dialysis process was first developed as a batch technique, using a simple dialysis flask fermentor [75,76]. This process is readily adapted to continuous culture (Fig. 13) [77].

In the simple continuous dialysis system shown, a culture of active yeast is

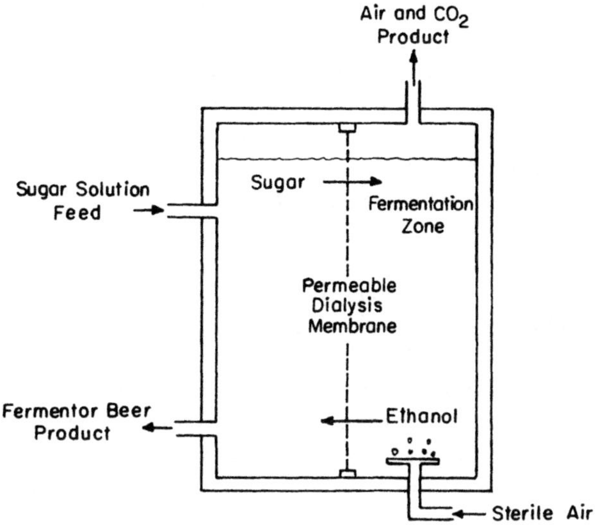

Fig. 13. Simple continuous dialysis fermentor; production rate limited by substrate diffusion to the fermentation zone

maintained in a confined zone of the vessel. Substrate enters the reaction zone by diffusing through a dialysis membrane from the medium zone. Reaction then takes place and product ethanol diffuses back through the membrane into the medium zone where it is recovered in an overflow. Yeast cannot escape the reaction zone and extremely high cell densities can be achieved.

For this system, reaction rate is limited by the rate at which substrate can diffuse across the limited membrane surface area, not by the inherent metabolic limitations discussed earlier. This practical diffusion limitation prevents very rapid reaction in

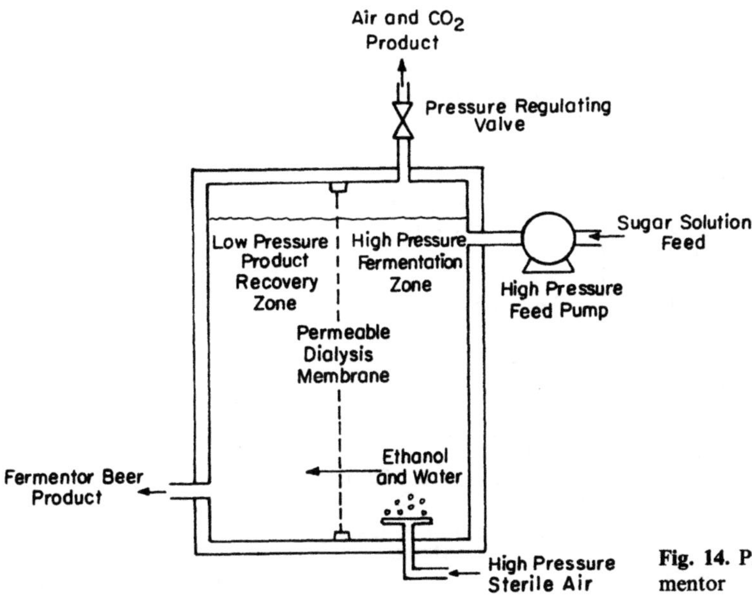

Fig. 14. Pressure dialysis fermentor

a simple system like that shown, and thus makes it impractical for industrial application to ethanol production.

The limitation of low substrate diffusion rate can be overcome if pressure dialysis is used (Fig. 14). Here, a pressure differential is applied across the dialysis membrane forcing a bulk flow of medium through a well-stirred culture zone. This system is also impractical because free proteins (the product of cell lysis) rapidly pile up on the membrane building up a thick cake which blocks the membrane pores and prevents further flow [78].

The problem of membrane fouling has been overcome in the rotorfermentor (Fig. 15) [58,79]. The rotorfermentor is a continuous pressure dialysis reactor. The simple fixed membrane is replaced by a rapidly rotating membrane cylinder. Feed and air are pumped into the annular reaction zone where yeast grow and are retained. This zone is maintained under pressure (115–170 kilopascals (2–20 psig)) [80], so that filtrate continuously flows out through the membrane. Because the membrane is rotating, a strong centrifugal force is developed at the membrane surface and large molecules impinging on the surface are thrown back into the annular zone. Only a very thin steady state cake thickness is developed. High filtration rates (3.4 $l\,h^{-1}$ for a 1020 cm^2 membrane surface, and 14 kilopascals (2 psig) pressure drop) can be maintained. Using a laboratory scale rotorfermentor cell densities of 50.9 $g\,l^{-1}$ and ethanol productivities of 36.5 $g\,l^{-1}\,h^{-1}$ have been achieved [80].

The rotorfermentor appears attractive in terms of its high productivity. It fails, however, when rated on mechanical and operating simplicity. The rotating membrane unit is mechanically complicated and seals for this unit will be made especially complex as they must not leak against the vessel pressure. More important though will be the requirement to periodically replace the membrane. The long term mechanical

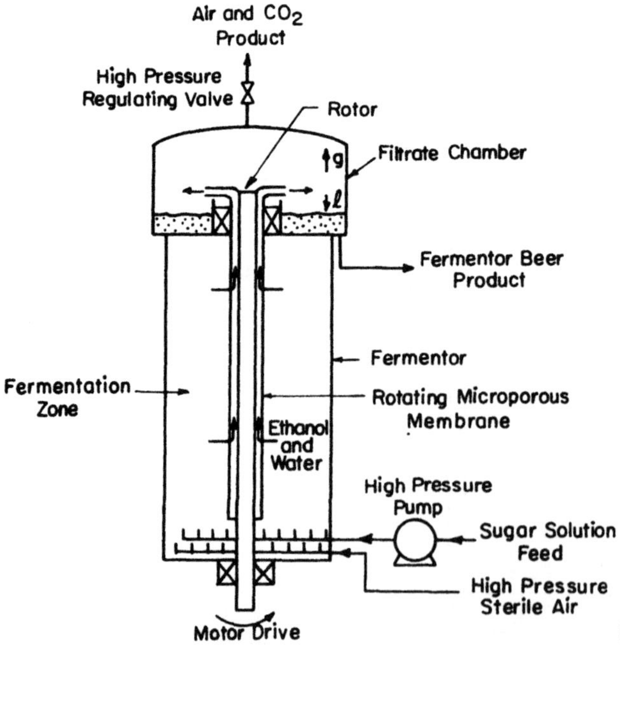

Fig. 15. Pressure rotorfermentor

stability of molecular filtration membranes under high shear has not been studied, but the developers of the rotorfermentor agree that periodic replacement will be necessary [81]. Membrane replacement will interrupt continuous operation and will require considerable skilled labor time. Because of these complications rotorfermentation probably will not become a competitive industrial process despite its high productivity.

Another approach to achieving high rate dialysis process involves the use of hollow fiber reactors (Fig. 16) [82]. Hollow fiber reactors provide extremely large membrane surface areas so that rapid substrate diffusion and high reaction rates are possible. The hollow fiber reactor is arranged like a shell and tube heat exchanger. The tubes are fine hollow fibers (0.05 cm I.D.) of membrane material [83]. A bundle of 1000 of these fiber tubes can be packed into a single three cm diameter shell [84]. The shell side is then inoculated with growing cells. Medium is fed into the fiber tubes at one end of the reactor and glucose substrate diffuses out and is converted to ethanol which diffuses back into the tubes and is carried out at the far end of the reactor.

Hollow fiber reactors have been used at laboratory scale to produce several high value biological products [82,85,86]. Work utilizing hollow fiber reactors for ethanol production is just beginning at Berkeley [87] and Stanford [88].

The hollow fiber reactor has potential to achieve yeast cell densities approaching the maximum cell packing density ($200\ g\ l^{-1}$) [89] and corresponding high ethanol productivities should be possible. Venting of carbon dioxide gas product may pose problems, however. Hollow fiber reactors are quite complex and very costly [84]. Membrane plugging may also be a problem. If these problems are not overcome, the hollow fiber reactor probably will not be important as an industrial ethanol reactor.

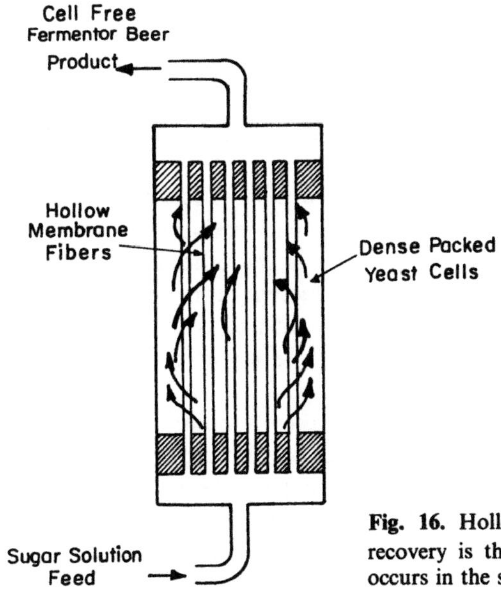

Fig. 16. Hollow fiber fermentor; feed flow and product recovery is through the fiber membrane bundle. Reaction occurs in the shell side high yeast density zone

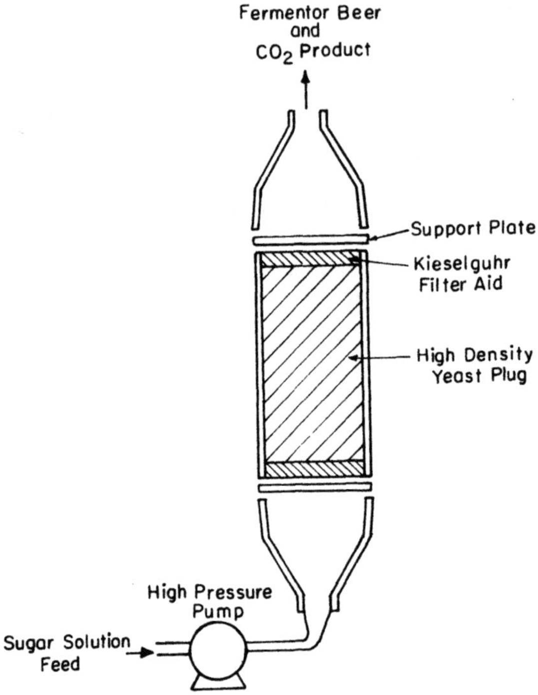

Fig. 17. High cell density plug fermentor

The plug fermentor (Fig. 17) is a particularly interesting version of the dialysis fermentor [90]. A dense plug of yeast is maintained between two support plates. Medium is pumped through the plug under high pressure (120–1100 kilopascals 0.2–10 atmospheres)) and rapid reaction takes place [91].

Easily fouled membrane filters are not used. Instead a layer of Kieselguhr filter aid over a porous frit support plate retains the yeast plug [42]. Kieselguhr is also mixed into the yeast plug to prevent dense packing of the yeast which would halt the flow.

Using this system, productivities 72 times greater than for simple batch with the same yeast and substrate have been achieved [90].

Oxygen concentration varies considerably across the plug. Yeast near the feed receive air salturated medium, but yeast near the product end receive oxygen depleted medium. Cell viability decreases with time because of this and other limitations and after 27 days operation, cell viability is only 52%. Productivity experiences a similar decline to 50% of the maximum level. Regular shutdown and rejuvenation of the yeast under ordinary growth conditions will probably be required for substained industrial operation.

The high productivity of the plug flow reactor makes it quite attractive. The requirement of high pressure equipment will partly offset the capital cost savings resulting from reduced vessel volume. High power pumps will be required to force medium through the plug, and the required shutdowns for yeast regeneration will interrupt continuous operation. Further study is required to determine if these disadvantages are offset by the very high productivities possible.

3.3.6 Cell Immobilization

Yeast cells can be immobilized by entrapment in a gel matrix or by covalent binding to surfaces [93].

Gel entrapment has been studied by many researchers [94,95]. In one study, yeast cells were suspended in 1% sodium alginate solution. This mix was then slowly extruded into .05 M $CaCl_2$ to produce fine fibers in which the yeast were entrapped. These fibers were then packed into a column reactor. 10% glucose solution was forced through the bed and reaction proceeded giving a 90% yield in 10 h residence time [94].

This system is especially simple as no agitation or yeast recovery equipment is required. It is found, however, that yeast viability declines for alginate gel entrapped cells with a cell half life of only 10 days [94]. Frequent gel fiber replacement is therefore required to maintain high productivity. Recent work using polyacrylamide gels suggests that the yeast viability problem can be overcome [96]. If high productivities can be maintained in an industrial gel-entrapment reactor, this could be a very attractive industrial process.

Another cell retention technique involves binding cells to solid surfaces. A thick gel can impose mass transfer limitations which slow the reaction rate [97]. These limitations are reduced if very thin gel layers on surfaces are used and are virtually eliminated if the cells are directly chemically bound to surfaces.

In one immobilization technique, ordinary distillation column packing is coated with a suspension of yeast cells in a polyelectrolyte. The polyelectrolyte is then crosslinked into a thin film gel with glutaraldehyde [98]. A laboratory column fermentor packed with this material achieved an ethanol productivity of 35 $g\ l^{-1}\ h^{-1}$.

Yeast cells have been bound to cotton gauze using cyanuric chloride to form a covalent bond between the exposed cotton hydroxyl groups and cell wall proteases or polysaccharides [97]. Ligand bonding to metal hydroxides has also been used [99]. Both of these techniques did result in cell retention and in both cases viability was maintained for long periods after immobilization. Further study is necessary to determine maximum productivities obtainable with chemically bound cells before any evaluation of the potential of this technique can be made.

3.3.7 Extractive Fermentation

The techniques described thus far have achieved high productivities by maintaining high cell densities. Now to be considered are techniques which can also achieve high specific ethanol productivities by maintaining low ethanol concentrations in the reactor and thus eliminating end product inhibition.

In the extractive process (Fig. 18) ethanol is continuously removed from the beer by solvent extraction [100]. A side stream of beer is tapped from the fermentor vessel, the cells are removed and recycled, and the clear beer is contacted with a liquid extractant. The extractant, which is immiscible in the beer absorbs most of ethanol. The purified beer can now be returned to the reaction vessel, and the ethanol saturated extractant can be processed through a distillation train for ethanol recovery and solvent recycle.

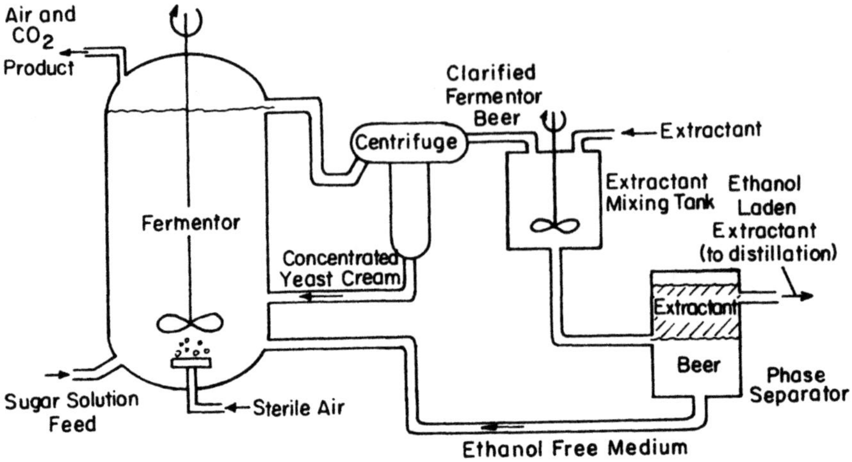

Fig. 18. Continuous solvent extraction fermentor

To be successful, an extractant must have these properties [101]:
1. non-toxic to yeast
2. high distribution coefficient for ethanol
3. selective for ethanol over water and secondary fermentation products
4. should not form emulsions with fermentation broth.

Extractive process has the potential to offer very high reaction rates by eliminating ethanol inhibition. If in addition to the required properties, the extractant is of low volatility, distillation will be simplified and considerable energy savings could result over conventional systems.

In terms of the process evaluation criteria, extractive fermentation, with a modest increase in complexity, yields a very high productivity, low-energy system. Unfortunately, no extractant having all the required properties has thus far been found [102].

3.3.8 Product Recovery Membrane Fermentation

Continuous ethanol recovery from the reaction mixture can also be achieved using selective membrane separation techniques.

Membrane extractive fermentation (Fig. 19) [102] is similar to simple extractive process except that the extractant is separated from the broth by a diffusion membrane. Ethanol, being more soluble in the extractant, diffuses across the membrane and is carried away. A membrane meterial which will allow ethanol removal while retaining sugar substrate in the reaction vessel can be used, and this helps assure complete substrate utilization [103].

Requirements for the extractant are far less severe for use in membrane extractive process than in direct extractive fermentation. Only a very small amount of extractant — that which leaks through the membrane — contacts the reaction mixture, and the requirements of nontoxicity can thus be reduced. The requirements of immiscibility and non-emulsion forming properties can also be reduced. The

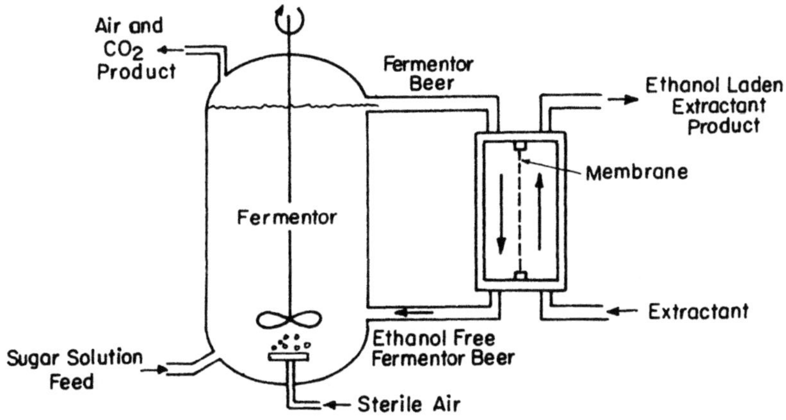

Fig. 19. Continuous membrane retractive fermentation

extractant polypropylene glycol p-1200 has been identified as suitable, having a distribution coefficient of 0.60 and being only slightly toxic to yeast [104].

Like simple extractive fermentation, membrane extractive fermentation appears promising. The process is simple, involves little added equipment and may again allow an energy reduction in distillation. Fouling of the membrane is a possible drawback as this would require frequent costly shutdowns for membrane replacement. Studies to determine maximum reaction rates achievable with membrane extractive fermentation are now underway [101].

Selective product recovery membrane fermentation is another new technique under development. The apparatus used is like that used for membrane extractive fermentation, except that no extractant is used. The membrane itself performs the separation, facilitating ethanol diffusion through the membrane while retarding water and other beer components. Selective ultrafiltration membranes capable of maintaining the ethanol concentration at less than 60 g l^{-1} while yielding a product concentration of 120 g l^{-1} have already been developed and membranes capable of maintaining beer concentrations at below 20 g l^{-1} should be available soon [105]. Flux rates across these membranes are still quite low and large membrane surfaces are required [106]. Membrane fouling may again be a problem. If these difficulties can be overcome so that continuous operation could be assured without frequent membrane replacement (so that membrane cost would not be prohibitive) then this process would be attractive for industrial use.

3.3.9 Vacuum Fermentation

The vacuferm process (Fig. 20) developed concurrently by Cysewski and Wilke [41] and Ramalingham and Finn [107] provides an alternative with continuous ethanol removal. A concentrated sugar solution is fed continuously to the reactor. Process is conducted under vacuum and an ethanol water solution is boiled away, maintaining the liquid level constant. Since ethanol is more volatile than water, this flashing operation acts as a single stage distillation. With a process temperature of 35 °C and pressure

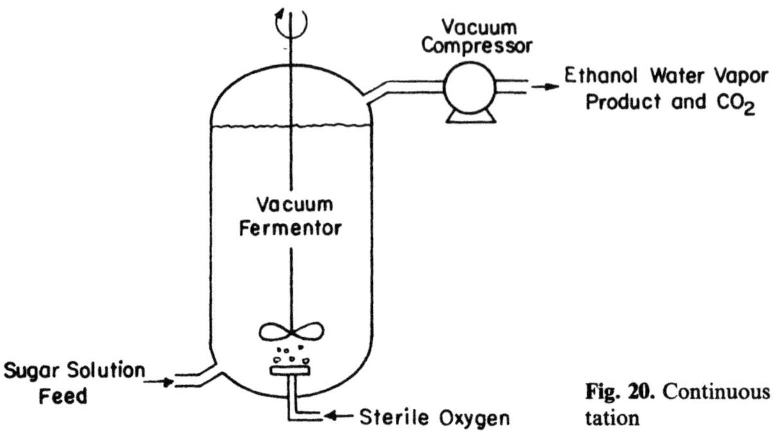

Fig. 20. Continuous vacuum fermentation

of 6.8 kilopascals (51 mm Hg), the ethanol concentration is maintained at 35 g l^{-1} while the ethanol concentration in the vapor product is increased to more than 200 g l^{-1}, thus simplifying later distillation steps [6]. Yeast cells build up in the fermentor and cell densities of 120 g l^{-1} can be maintained. With ethanol inhibition removed a total productivity of 80 g l^{-1} h^{-1} can be achieved [61].

There has been some concern over the added energy requirements to drive vacuum compressors for the vacuferm process [62]. It has been shown, however, that energy requirements for the vacuferm process are increased only five percent over those for conventional processes when suitable techniques for energy recovery are amployed [108].

The absolute productivity achieved in the vacuum fermentor is higher than has yet been achieved in other devices and this advantage far outweighs the small increase in energy requirements for vacuum compressor operation. Other difficulties are associated with vacuferm operation, however. To meet the yeast oxygen maintenance requirements under vacuum, pure oxygen must be sparged, and this contributes an added cost of 0.5 cents per l the cost of alcohol produced [6]. The compressors required must operate at unusually low-pressures and are extremely large. They will be difficult to control [109]. Capital cost of the compressors appears to have been underestimated in published evaluations of this process and may actually more than offset the savings in reactor vessel cost. Vacuum operation will increase the likelihood of fermentor contamination and shutdown. It is not clear whether the outstanding productivity of the vacuum fermentor can compensate for these potential difficulties. Pilot plant operation has been proposed [110], however, and should resolve this question.

The flashferm process modifies the simple vacuferm process to overcome some of its operating diffisulties (Fig. 21) [6]. Reaction is carried out in an atmospheric pressure fermentor so that the yeast oxygen maintenance requirement can be cheaply met with sparged air. Carbon dioxide produced can be directly vented from the vessel and thus needs not be compressed with the vapor product as in the vacuferm process. To remove ethanol, beer is rapidly cycled through a small auxiliary flash vessel where

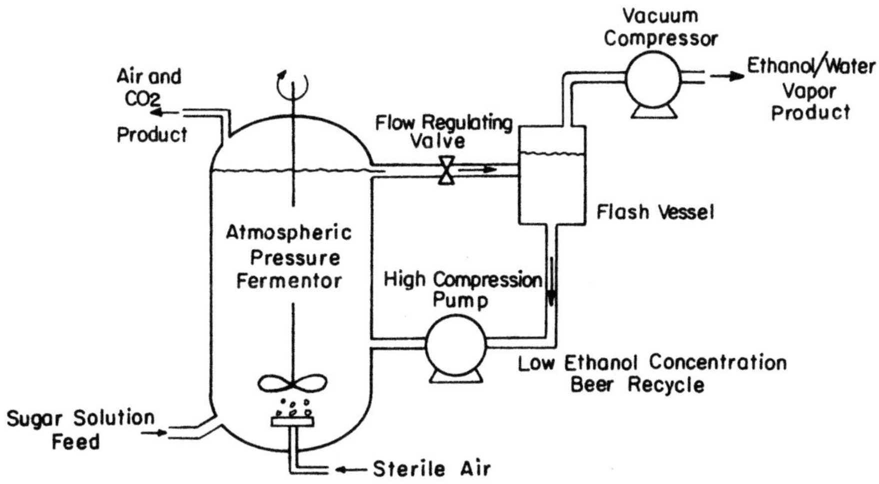

Fig. 21. Continuous flash fermentation

it boils. Ethanol is recovered as the flash vessel overhead vapor product and ethanol depleted beer is returned to the vessel. The problem of contamination is greatly reduced as only the small flash vessel is under vacuum. Energy requirements for this process are also slightly lower than for vacuferm.

The flashferm plant is somewhat more complicated than the vacuferm plant, the former requiring an added vessel and the associated liquid cycling pumps, but this is probably more than compensated by the advantages cited. Like the direct vacuferm process, it is not clear whether the high productivities possible with vacuum operation offset the operation and control difficulties and the likely high compressor costs.

3.3.10 Discussion

The twenty different advanced high rate processes described in the proceeding section have been assessed as to their potential for application in industrial alcohol production. The comparisons made are summarized in Table 6. Clearly many alternatives superior to the conventional batch technology of the 1930s exist. Among these, simple continuous, series continuous, cell recycle and APV tower fermentors have been operated at large scale with considerable savings over batch processes. These should then be the processes of choice for any new ethanol plants.

Among processes with continuous ethanol removal to eliminate end product inhibition, only vacuum and flash fermentation have been advanced sufficiently to allow pilot plant testing.

The other high rate processes, dialysis bound cell and extractive and selective membrane processes require much more evaluation before their merits can be fully assessed.

Table 6. Alternative fermentation processes

Process	Ethanol productivity	Comments	Ref.
Batch fermentation	Very low ($1.8–2.5\ g\ l^{-1}\ h^{-1}$)	Very high capital and operating cost	3, 5)
Simple CSTR fermentor	Low ($6\ g\ l^{-1}\ h^{-1}$)	Mechanically simple equipment, simple continuous operation	5, 41, 50)
Series CSTR fermentors	2–3 times rate for simple CSTR	Simple continuous operation	2, 33, 47, 48, 51)
Perforated plate	Not reported	Mechanically complex	52, 53, 54, 55, 56)
CSTR with centrifuge cell recycle	High ($30–40\ g\ l^{-1}\ h^{-1}$)	Added energy requirements and added operator attention required for centrifuge	33, 41, 44, 62, 57)
CSTR with alternative recycle scheme	High	Settlers appear to be too large for industrial application. Whirlpool separators may be attractive	63, 64, 65, 66, 67)
Tower fermentors	High	Mechanically simple with simple continuous operation, but start up period is very long	2, 68, 69, 70, 71, 72)
Slant tube fermentors	High	Mechanically complex	73)
Packed bed, tower fermentor	High	Plugging and by-passing are major problems	74)
Dialysis fermentor	Not reported	Reaction rate limited by substrate diffusion through membrane. Membrane fouling may require frequent shutdowns	75, 76, 77)
Pressure dialysis fermentor	Not reported but potentially high	Membrane fouling will be a major problem	58, 78)
Rotor fermentor	High ($36\ g\ l^{-1}\ h^{-1}$)	Membrane fouling problem overcome but the fermentor device is mechanically quite complex. Membrane destruction by mechanical shear may be a problem	58, 79, 80)
Hollow fiber	Not reported but potentially high	Expensive units. Mass transfer limitation may limit productivity	82, 83, 84, 85, 86)
Plug fermentor	High (72 times greater than for similar batch fermentation)	Frequent shutdowns for yeast rejuvenation in an aerobic environment will be required. High power feed pumps are required	90, 91, 92)
Gel entrapment	90% yield from 10% glucose solution in 10 h	Very simple system with no agitation or recycle equipment required. Half life of entrapped cells must be increased to make this process attractive	94, 95, 96, 98)
Chemically bound yeast cells	Not reported	The cell viabilities do remain high after covalent bonding. If productivities also remain high, this could yield in very simple, high productivity system	93, 97, 99)

Direct extractive fermentation	Potentially very high	No suitable extractant has been found	100, 101, 102)
Membrane extractive fermentation	Not reported, but potentially very high	Membrane fouling may be a problem. Process is otherwise simple	102, 103, 104)
Selective membrane fermentation	Not reported but potentially very high	Membranes with sufficiently high throughput rates have not yet been developed	105, 106)
Vacuum fermentation	Very high ($80\ g\ l^{-1}\ h^{-1}$)	Mechanically complicated equipment requiring constant monitoring. Small added energy requirements for vapor compression. Contamination of the vacuum vessel may be a problem. Pure oxygen must be sparged	41, 61, 62, 107, 108, 109, 110)
Flash fermentation	Very high	Mechanically very complicated. Energy requirements only slightly increased over conventional processes. Contamination problem greatly reduced as compared to direct vacuum process and pure oxygen is not required	6)

4 Alcohol Product Purification

4.1 Introduction

Alcohol concentration and purification is an expensive step in industrial alcohol production as distillation is a large energy consumer. The use of bioprocess alcohol as a liquid fuel has been questioned on the basis that the energy required for distillation is equal to the total combustion energy of the alcohol product (21.1×10^6 Joules per l) [111, 112]. While inefficient beverage alcohol plants and plants producing ultrapure ethanol utilize nearly this energy, well-designed alcohol plants of the 1940s utilized as little as 2.5 kg of steam per l of product (about 5.57×10^6 Joules) for distillation. Extending the vapor reuse techniques of the 1940s, some modern grain plants are able to cascade cooking, washing, distillation and stillage drying operations (reutilizing steam from each step into the next) to produce industrial alcohol with less than 3.0 kg of steam per l in total [59, 113, 169, 170]. Further, new processes now under development may improve distillation or replace it entirely with more efficient separation processes.

4.2 Conventional Distillation Technology

4.2.1 Background

The separation of dilute alcohol-water mixtures into pure alcohol and pure water is a difficult problem as alcohol and water form an azeotrope at 95.7 wt% (89 mole%) alcohol concentration [114]. Figure 22 shows the relative volatilities of ethanol and water [115]. From 0 to 89 mole% alcohol, ethanol is more volatile than water-

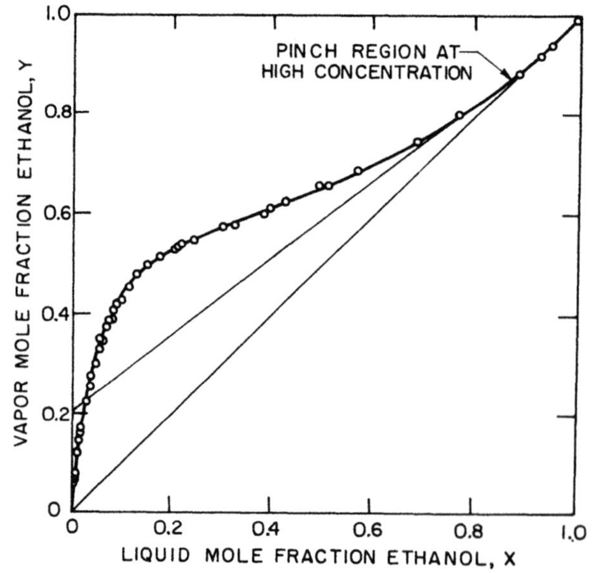

Fig. 22. Atmospheric pressure ethanol/water equilibrium

concentrating in the vapor phase. As the azeotrope composition is approached from below the relative volatility of alcohol over water decreases until, at the azeotrope, the volatilities of ethanol and water become identical. Boiling gives a vapor of the same composition as the boiling liquid and no further separation can be achieved by simple distillation. An additional chemical agent must be added to alter the equilibrium and allow a separation. Thus, to produce pure alcohol, two stages of distillation (one to produce the azeotrope and the second, with an added chemical agent to alter the equilibrium) are required.

The preparation of industrial alcohol from beer is further complicated as the reaction mixture contains many by-products [138)]. Most important among these are aldehydes and fusel oils. The distillation system must be designed to remove these by-products.

Typically, 1 l of acetaldehyde (and much smaller amounts of other aldehydes) is produced for every 1000 l of alcohol product [116,117)]. Acetaldehyde is very volatile — with a boiling point of 21 °C, and thus, acetaldehyde is easily separated from the alcohol water mixture as a distillation head product.

As much as 5 l of fusel oil (see Table 7) [10,119,120)] are produced for each 1000 l of alcohol produced. This mixture of amyl and propyl alcohols and their isomers exhibits unusual equilibrium behavior which complicates its separation from the alcohol water mixture. Dry fusel oils boil in the range from 128–137 °C. Dry fusel oils are much less volatile than ethanol and would not be expected as a distillation head product. Fusel oils also are not entirely miscible with water. The volatility of fusel oils is greatly enhanced in the presence of water with only 2.4% amyl alcohol in water exerting the same partial pressure as 90.7% amyl alcohol in 9.3% water [122)]. The greatly enhanced vapor pressure of dilute amyl alcohol in water forces amyl alcohol into the vapor phase near the bottom of an alcohol purifying column and fusel oils thus cannot be separated as a distillation bottoms product. Fusel oils concentrate in the middle of an alcohol purifying column, building up until they finally "leak" out as contaminants of both the head alcohol and bottoms water streams.

To remove the fusel oils and prevent product contamination, fusel oil laden liquid is bled from the lower plates of the alcohol rectifying section. This fusel oil is cooled and allowed to separate, then contacted with clean water to recover alcohol for recycle to the main still.

4.2.2 Conventional Distillation Processes to Produce Azeotropic Alcohol

With the basic distillation problem outlined, some of the many alternative distillation schemes which are practiced for ethanol purification can be considered.

Table 7. Fusel oil yields

Substrate	Fusel oil; l per 1000 l alcohol	Substrate	Fusel oil; l per 1000 l alcohol
Blackstrap molasses	4–5	Degerminated corn	2.5–3.0
High test molasses	1–3	Wheat	2–3
Corn	4–5	Rye	2–4

The simplest systems utilize only alcohol stripping and rectifying sections (as either one or two columns) with fusel oil tapped from the bottom half of the rectifier (Fig. 23). Beer (normally 6 wt % alcohol) is fed to the top of the stripping section (or beer still). Steam fed to the still bottom strips alcohol, aldehydes and fusel oil from the descending beer liquid to give a purified water bottoms product (containing also stillage solids). Aldehyde — being most volatile — is the major product taken from the rectifying column head. Aldehydes are condensed at the head to provide reflux. A few trays below the head, an alcohol/water liquid draw — containing some aldehyde — is taken. This is the industrial azeotropic alcohol product. Fusel oils are removed as a bleed from the rectifying section lower plates, cooled and water washed for dilute alcohol recovery.

This design involves the simplest equipment, but products are of low quality. The alcohol product must necessarily be contaminated with substantial amounts of aldehyde. 2.4–3.0 kg of steam are required per l of product.

For higher alcohol quality, a three column design is used. (Fig. 24) [148,149]. Extensive heat recovery allows very efficient operation. As in the simpler unit discussed, beer is first fed to the head of the beer still and alcohol is stripped from the descending beer with steam. The alcohol concentrated vapor stream (50 wt % alcohol) is then divided. The first portion is fed to the reboiler of an aldehyde stripping column and condensed to provide boil-up heat for this still. The remainder of the vapors are fed to a rectifying column where further water is removed. High alcohol liquid with some aldehydes and fusel oil is drained from near the head of the rectifying column and fed to the aldehyde stripping column. Fusel oils are bled at a lower plate side tap and the aldehydes are stripped out to give an essentially pure azeotropic alcohol-water mixture at the bottom of this column. The

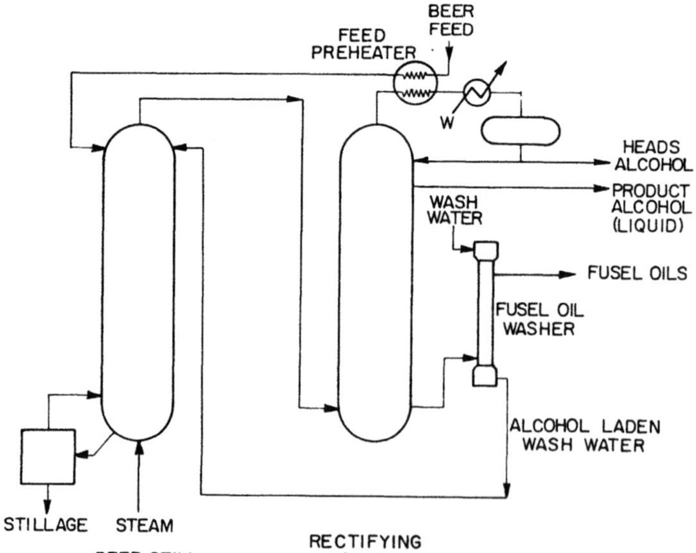

Fig. 23. Two column ethanol/water distillation system

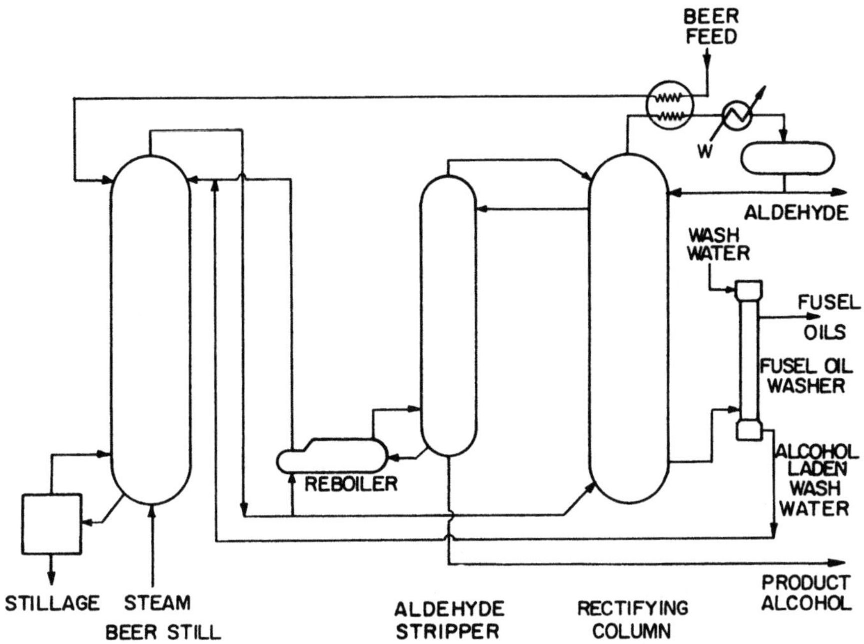

Fig. 24. Three column ethanol/water distillation system

energy requirement is identical to the simple one (or two) column design — 2.4–3.0 kg pounds of steam per l of product.

A four column "Barbet" design which incorporated a fusel oil concentrating column was common. This design offers no advantage over the three column design, however. Product qualities are not enhanced and energy requirements are higher (4.1 kg of steam per l of product) [10].

A final design of interest utilizes six columns to produce azeotropic ethanol and aldehydes of high purity. The beer still is operated at low pressure so that boiling temperatures can be reduced. This prevents the breakdown of yeast in the stillage which otherwise would produce trace organic contaminants in the final alcohol product. This complex system is necessary only when high purity reagent alcohol is desired.

4.2.3 Ethanol Dehydration — Conventional Processes

The ethanol water equilibrium must be altered to allow production of alcohol drier than the azeotropic composition by distillation. Anhydrous ethanol is most usually produced by azeotropic distillation with an added entrainer.

The benzene azeotropic distillation is best known (Fig. 25) [123, 124]. In the azeotropic distillation, benzene acts as an entrainer. Water is immiscible with highly non-polar benzene and hence, the vapor pressure of water above a boiling benzene/ethanol/water liquid mixture is greatly enhanced. A rich phase (74 mole% benzene decanted from the cooled distillate) is used as reflux to the head of the azeotropic column and flows downward as a liquid. The azeotropic ethanol/water mixture is fed near the

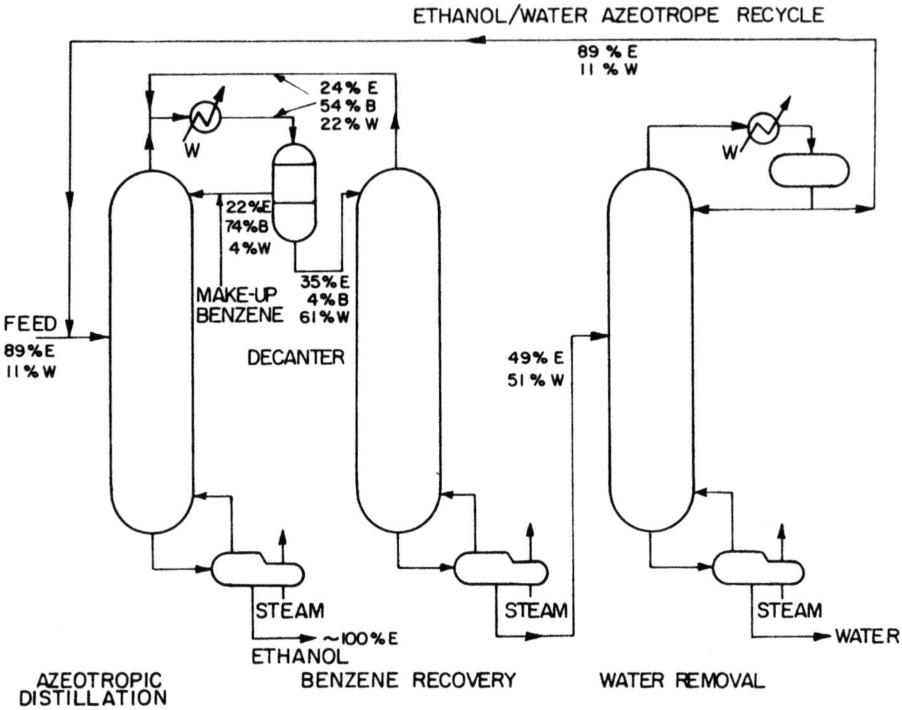

Fig. 25. Azeotropic distillation with benzene for pure ethanol production

middle of the column. Above the feed, the volatility of water is increased by the presence of benzene and the ethanol volatility is relatively depressed. Water is concentrated in the vapor and a water concentrated head product (24% ethanol, 54% benzene, 22% water) results. This is the ethanol/benzene/water ternary azeotrope. Below the feed tray, benzene is stripped from the ethanol to give a pure ethanol bottoms product. When the distillate of the azeotropic column is cooled and decanted, a water rich phase (35% ethanol, 4% benzene, 61% water) is removed. This is treated in a benzene recovery column which produces the benzene concentrated ternary azeotrope as overhead product for recycle to the decanter and a dilute ethanol bottoms product (49% ethanol, 51% water). The dilute ethanol stream can be concentrated in an additional rectifying column or recycled to the primary alcohol recovery column.

Production of anhydrous alcohol from the azeotrope using benzene as an entrainer requires 1.0 kg of steam per l of product. The distillation columns are quite costly, as the azeotropic column alone requires 50 plates [126].

Other entrainers have been used. Both the trichloroethylene "Darwinol" process and the n-pentane process operate similarly to the more common benzene process [125]. These require smaller columns and have a slightly reduced energy input [126].

Ether offers a major advantage as an entrainer in that a ternary ethanol/water/ether azeotrope is not formed and the azeotropic column head product is an alcohol free, ether/water mixture [127,128,129,130]. Thus, all the alcohol entering the azeotropic column as feed is removed as bottom product and there is no energy inefficient recycling of alcohol/water mixtures.

As in the benzene process, the azeotropic column head product (3.8% water) is cooled and decanted, giving an ether phase (98.8% ether) for reflux and a water rich phase which must be stripped of ether in an auxiliary column. Ether and water are very far apart in boiling points and the ether recovery column requires very few plates.

The steam requirement of the azeotropic column is about 0.42 kg per l of alcohol product. This column, however, is generally operated at from 690–830 kilopascals (100–120 psig) with a head temperature above 100 °C so that condensing head vapors can provide heat for other columns in the distillation train. Using this method, and teaming the ether distillation with the three column vapor-reuse process, anhydrous alcohol can be produced from 6 wt% fermentor beer using only 2.5 kg of steam per l of product [127].

Extractive distillation is the major alternative to azeotropic distillation. In extractive distillation, a chemical agent is added which breaks the azeotrope by enhancing the volatility of the ethanol. Pure ethanol is then taken as the head product while a mixture of water and the extractant emerges as the bottom product. This process offers the advantage that separate columns are not required to produce the azeotrope and then produce pure alcohol. Instead a single rectifying column is used with the extractant added above the feed. An auxiliary column is required to recover extractant from the water for recycle. Glycerol is an effective extractant, but this process has not been widely used [132,133].

4.3 Alternative Processes for Alcohol Purification

As can be seen from the above review of conventional technology, distillation processes are well advanced. New processes are under development, however, which may reduce the cost and energy requirements for alcohol purification.

4.3.1 Vacuum Distillation

The ethanol/water equilibrium is pressure sensitive, and the relative volatility of ethanol over water at high ethanol concentration is increased at low pressure (Fig. 26). As pressure is reduced from one atmosphere, the ethanol water azeotropic composition increases until, at about 10.6 kilopascals (80 mm Hg) absolute pressure, the azeotrope disappears. Pure alcohol can thus be distilled from an alcohol water solution in a single step at reduced pressure. The energy required for this "one-step" distillation is very high and this process is uneconomical. However, vacuum operation can be used to advantage in producing a product of the normal (atmospheric pressure) ethanol/water azeotropic composition. Under vacuum, the required reflux may be reduced. For a 13 wt% alcohol feed (no aldehydes or fusel oil) producing 95 wt% alcohol product, the energy requirement for a single column distillation is reduced from 7.00×10^6 Joules l^{-1} of product to 2.41×10^6 Joules l^{-1} [6].

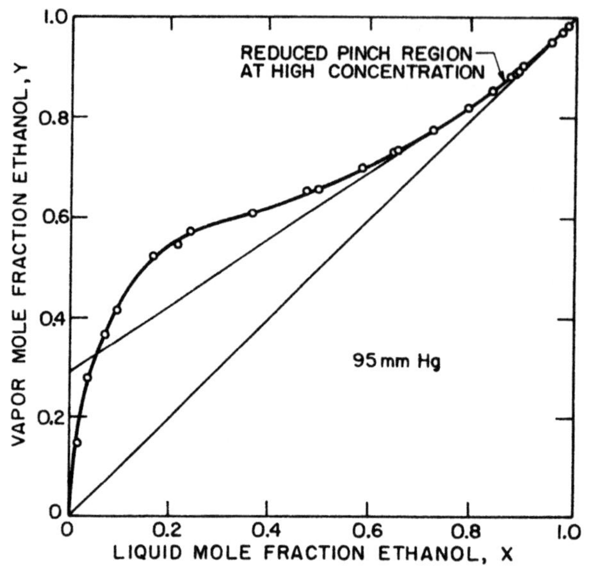

Fig. 26. Low pressure ethanol/water equilibrium

4.3.2 Extractive Distillation-Salt Effect

Furter has proposed using potassium acetate or other salts as an extracting agent for ethanol distillation [135, 136, 137]. Potassium acetate greatly enhances the relative volatility of ethanol over water, completely eliminating the azeotrope and greatly reducing the number of stages and the energy required for the distillation [141]. The column is operated with salt added to the reflux. Pure alcohol is the head product, but the bottoms water product contains salt which must be separated for recycle. Salt recovery by evaporation of the water is impractical, but application of technology developed for ocean water desalination may make this process more attractive.

4.3.3 Ethanol Dehydration by Vapor Phase Water Absorption

Dehydration of azeotropic ethanol vapor by absorbing water onto dry calcium oxide or potassium carbonate is a standard laboratory procedure [139, 140]. Ladisch and Dyck [140] suggest that similar processes may be used beneficially at industrial scale. Rather than distilling to the azeotrope, distillation is conducted to produce only 85 wt% alcohol vapor. Distillation to this lower level requires only 1.73×10^6 Joules l^{-1} for an 85% ethanol product. These vapors are then passed through a bed of adsorbent material to produce pure alcohol. Laboratory scale tests indicate that dry corn starch will selectively adsorb all of the remaining water and essentially no alcohol. This process seems quite attractive, but further work appears necessary to determine the energy requirement for regeneration of the absorbent and to test the process at pilot plant scale.

4.3.4 Solvent Extraction

Solvent extraction offers an alternative to conventional distillation. Myers suggests countercurrent extraction of ethanol from water solution into diethyl ether [143]. The high volatility of ether and its low heat of vaporization would allow a simple low energy distillation. Separation delete factors can be increased and extraction made more efficient by the addition of salts which increase the activity of the ethanol in the aqueous phase and help drive it into the organic solvent [144]. Such a process has been proposed for the direct production of gasohol by extraction of ethanol from beer into gasoline [145, 146]. As with extractative distillation using salts, a low energy technique must be developed for salt recovery from the aqueous phase.

4.3.5 Exotic Recovery Processes

Several more exotic processes for ethanol purification have been proposed. Filippi suggests alcohol extraction with supercritical carbon dioxide at 8000 kilopascals (80 atmospheres), followed by CO_2 flashing for ethanol recovery [147]. Membranes which will pass water while retaining alcohol have been developed, but the extremely high pressure driving forces required and the low throughput rates (flux) achieved make industrial application impractical [105, 106, 147]. Selective absorption of water into zeolite molecular sieve cages [150] has also been suggested [147, 151]. These, and other such exotic processes, may someday become workable but much more research is required before their potential merits can be evaluated.

4.4 Ethanol Purification Process Comparison

Table 8 compares the various ethanol purification processes. Unlike conventional ethanol processes, conventional ethanol distillation practice is well advanced. The three-column vapor reuse distillation combined with an ether or pentane azeotropic distillation to produce anhydrous alcohol is quite efficient and is probably the process of choice for new ethanol plants. Of the new proposed processes, vacuum rectification and vapor phase water adsorption are ready for pilot plant testing. Liquid/liquid extraction has great potential for energy savings, but considerable research is required to find a suitable solvent.

5 By-Products

By-products can be a major revenue source, important in achieving a positive return on alcohol production [3, 10, 152, 153, 154, 155]. Major by-products are stillage, fusel oil and carbon dioxide. With cellulosic raw materials, non-convertible pentose sugars, furfural, and lignin are also produced [15].

Table 8.

Process	Conventional processes	Comments	Ref.
	Energy consumption		
Simple 1 or 2 column distillation to 96 vol% ethanol from 6.25%	2.4 kg steam per l	20 tray beer still 30 tray rectifying column	10)
3 column vapor reuse process for high quality 96 wt% industrial alcohol from 6.25 vol%	2.4 kg steam per l	20 tray beer still 45 tray aldehyde column 54 tray rectifier	148, 149)
4 column Barbet unit for 96 vol% industrial alcohol from 6.25 vol%	4.1 kg steam per l	20 tray beer still 30 tray purifying column 54 tray rectifying column	10)
Benzene dehydration from 96 to 99.9 vol% ethanol	1.0 kg steam per l	50 tray dehydrating column 30 tray water-steam column 45 tray supplementary rectifier	123, 124, 126)
Combined 3 column vapor reuse and high pressure ether distillation for 99.9% ethanol from 6.25%	2.5 kg steam per l	20 tray beer still 45 tray aldehyde column 54 tray rectifier 30 tray dehydrator 20 tray water stripper	127, 128, 129, 130)
Pentane dehydration from 96 to 99.9 vol% ethanol	0.9 kg steam per l	18 tray dehydration column 14 tray water stripper column supplementary rectifier also required	125, 126)
Vacuum rectification to produce 95 wt% ethanol from 13 wt%	1.1 kg steam per l	40 tray rectifier (This design assumes no aldehydes or fusel oil)	6, 134)
Extractive distillation using salts to produce 99.9 mole% ethanol from 5wt% ethanol	1.1 kg steam per l for distillation, additional energy for salt recovery depending on the process chosen	Single 20 tray column (This design assumes no aldehydes or fusel oil) Success of this process hinges on low-energy recovery of salts, not yet studied	135, 136, 137, 141)
Vapor phase adsorbtion of water to produce 99.9% alcohol from 85% alcohol	Energy requirements for absorbent, regeneration have not been established	only small laboratory tests have been conducted	139, 140, 141)
Extraction	Laboratory scale process requires further study		143, 144, 145, 146)
Membrane separation	Laboratory scale process		147, 150, 151)
Molecular sieve separation	Conceptual process requires further study		105, 106, 147)

5.1 Stillage

Stillage is the sum of all insoluble and non-volatile matter remaining after the process. Stillage is recovered from the base of the beer still in the beer still bottoms product. Screening or centrifuging recovers distillers light grains — high protein yeast, and solid residue from the fermentor feed (such as corn residue). Multi-effect evaporation of the remaining liquid gives a dark syrup — distillers solubles, which can be mixed back with the light grains to give dark grains cattle feed.

The vitamin and protein content of stillage varies considerably with the source. A typical analysis for corn stillage is given in Table 9 [10]. An evaluation of the stillage by-product credit (cent per l ethanol) is given in Table 10 [152] for several sugar sources. Cattle feeding studies with stillage from cellulosic raw materials are needed before a cellulosic stillage credit can be established.

Table 9. Nutrient content of corn stillage

Vitamin or carotenoid	Vitamin content, p.p.m., in:		
	Solubles	Dark grain	Light grain
Thiamine (B_1)	6–10	3–4	1.2
Riboflavin (B_2)	15–30	7–10	2–3
Pantothenic acid	21–36	10–13	3–4
Niacin	90–160	40–9	15–30
Pyridoxine (B_6)	8–10	—	—
Biotin	0.4–0.5	0.8–0.22	0.04
Choline	6000–7000	4000–5000	1500–2000
Carotene	0.8	1.0–1.2	1.8
p-Aminobenzoic acid	9–10	—	—
Folic acid	4.0	—	—
Zeaxanthin	6.5–8.8	7.5–8.3	8.0
Cryptoxanthin	3.8	5.0–5.6	5.1
Protein		30 wt%	30 wt%

5.2 Fusel Oil

Fusel oils are produced by biodegradation of amino acids in the fermentation raw material [150, 157, 158, 159]. Fermentation of reagent glucose gives only traces of fusel oils. Standard sugar sources may yield up to 5 gallons of fusel oil for every 1000 l of ethanol (Table 7) [10].

Fusel oil is a mixture of amyl and propyl alcohols and their isomers (Table 11) [10, 160] — similar to major gasoline components, but with an added alcohol group. As such, the major use of fusel oils from an ethanol fuel plant would be as further gasoline supplement and fusel oils should be valued at roughly the value of gasoline. Alternatively, fusel oils can be sold as a solvent or tanning compound at 4.5 cents per kg.

Table 10. Stillage credits

Item	Units	Corn	Grain sorghum	Wheat	Cull potatoes	Sugar beets	Molasses	Cellulosics
By-product selling price	dollars per t	121	121	121	7	102	17	Unknown
By-product yield	kg by-product	7.63/bu	7.63/bu	9.41/bu	135/t	120/t	7.09 l^{-1}	
Unit by-product credit	cents per l ethanol	9.5	9.5	11.6	0.8	15.9	3.2	

* Potato and molasses distillers by-products 75% moisture, others dry

Table 11. Composition of fusel oil

Constituent	Composition of fusel oil, % constituent from:			
	Corn	Molasses	Potatoes	Rye
Isopropyl alcohol		0.6	—	—
n-Propyl alcohol	20.4–11.7	24.3	6.85	—
Isobutyl alcohol	23.9–12.2	7.4	24.35	15.7
n-Butyl alcohol	—	8.1	—	—
active-Amyl alcohol	14.6–23.4	55.3	68.76	79.8
Isoamyl alcohol	36.3–59.7			
n-Amyl alcohol	—	4.3	—	—
Undetermined	4.8–3.0	—	0.04	6.5

5.3 Carbon Dioxide

Carbon dioxide is produced in a one to one molar ratio with ethanol. Compression to produce dry ice is possible, but a very large market would be required near the plant. Even then, the return would be small.

An alternative use for carbon dioxide is in methanol synthesis [161]. On site methanol generation was practiced in some WWII alcohol plants [10] and this may become an attractive by-product use again.

5.4 By-Products from Cellulosic Feeds

Preparation of sugar from cellulosic feed will generally give substantial amounts of pentose sugar and lignin by-products [162,163]. Corn stover, for example should yield 0.37 kg xylose and 0.43 kg lignin per kg of degradable sugar (based on stover composition and 100% conversion). Xylose could be recovered as part of the stillage and contribution to cattle feed quality.

Lignin is now used only as a low grade fuel, with a heating value of 29.5×10^6 Joules per kg [164] and a market price of $ 110 per ton [165]. Use of lignin and furfural (a xylose degradation product produced in acid hydrolysis) for plastics manufacture has been proposed [166]. Lignin may also be used as a lignin-formaldehyde binder, replacing phenol-HCHO in particle board binders [165]. This could result in a substantial by-product credit, but much more research is required.

6 Ethanol Production Economics

Ethanol cost is established primarily by selection of the sugar source. Plant design and capacity, energy requirements and fuel cost, by-product credits, and form of financing also have significant effects. Economic evaluations for many design cases are available in the literature [95,6,152,153,167]. An evaluation by Raphel Katzen Assoc. for the U.S. Department of Energy [7] is especially complete. The Katzen design utilizes corn as the sugar source and corn is likely to be the most important substrate in the U.S. gasohol program in the near term.

6.1 Corn to Alcohol Plant Design Basis

200 million l per year industrial ethanol from corn substrate plant will be considered. Continuous mashing cooking and saccharification with a fungal amylase enzyme are used, followed by traditional batch reactions in 380,000 l vessels. Distillation is by an efficient vapor reuse process with steam generated on site from coal.

Feed inputs and products from the plant are summarized in Table 12 [7].

Table 12. 200 million l per year corn alcohol plant: Material flows

Inputs	Rate
Corn	67,500 kg h^{-1}
Coal	11.3 t h^{-1}
Yeast	45 kg h^{-1}
Denaturant	238 l h^{-1}
Ammonia	349 kg h^{-1}
(for coal flue gas purification)	
Products	
199 Proof motor fuel grade ethanol	23,900 l h^{-1}
By-products	
1. Distiller's dark grains	19,900 kg h^{-1}
2. 40 wt% $(NH_4)_2SO_4$ solution	2,990 kg h^{-1}
(from coal flue gas purification)	
3. Fusel oil	87.2 kg h^{-1}

6.2 Plant Energy Requirements

Energy costs will be an increasingly important factor in ethanol production economics as fuel costs rise. The total energy requirement is 90,000 kg of steam per h and 8,314 kw. Efficient vapor reuse is employed throughout the plant. The distribution of energy use between the various process steps is summarized in Table 13.

Coal for steam is purchased at 26.95 per ton and electricity is purchased at $ .03/kw.

6.3 Plant Capital Costs

Capital cost for the 200 million l per year alcohol plant are shown by plant section in Table 14 [7]. The total plant cost (Dec. 1978 basis) is $ 57,991,000.

Table 13. Process Steam and Electrical Power Breakdown 200 Million l per year Alcohol Plant (from Katzen et al. [7])

Process section	Process steam	Electrical
Total plant requirement	90,000 kg h^{-1}	8,314 kw
Receiving, storage and milling	0.0	6.1%
Mash cooking saccharification	30.5%	26%
Fungal amylase production	0.7%	20.4%
Bioconversion	0.2	4.0%
Distillation	58.5%	1.6%
DDG Recovery	6.4%	27.1%
Storage denaturing	0.0%	0.7%
Utilities	2.7%	37.0%
Buildings	1.0%	0.5%
Total	100.0%	100.0%

Table 14. 200 million l per year Plant Investment Katzen, et al. [7)]

Process section	December 1979 cost ($)	% of investment
Receiving, storage milling	2,086,800	3.96
Cooking and saccharification	2,824,300	5.36
Fungal amylase production	3,485,900	6.61
Fermentation	4,195,600	7.96
Distillation	5,123,800	9.72
Dark grain recovery	13,018,400	24.69
Alcohol storage, denaturing and by-product storage	4,399,900	8.34
Utilities	15,090,000	28.62
Building, general services and land	2,494,000	4.73
	52,719,000	100.00
+10% Contingency	5,272,000	
Total plant cost	$ 57,991,000	

6.4 Alcohol Manufacturing Costs

Manufacturing costs (Dec. 1978 basis) are summarized in Table 15 [7)]. After allowing reasonable by-product credits, a total operating cost of 23.5 cents per l results. The alcohol selling price required for a 15% discounted return is then $ 27.7 per l [7)].

Table 15. 200 million l per year Plant Operating Cost

	Equivalent cost cents per l		Equivalent cost cents per l
Fixed charges		*Labor*	
Depreciation*	1.53	Management	0.13
License fees	0.03	Supervisors/operators	1.16
Maintenance	0.95	Office & laborers	0.63
Tax and insurance	0.50	Subtotal	1.92
Subtotal	3.01	Total production cost, TPC	31.20
Raw materials		*By-products*	
Yeast	0.16	Dark grains	10.15
NH_3	0.18	Ammonium sulfate	—.24
Corn	23.67	Subtotal	−10.39
Coal	1.27	*Miscellaneous expenses*	
Miscellaneous chemicals	0.11	Freight	1.32
Subtotal	25.39	Sales	1.03
Utilities		G & AO	—.34
Electric power	0.87	Subtotal	2.69
Diesel fuel	—	*Total operating cost*	23.51
Steam (from plant)	—		
C.W. (from plant)	—		
Subtotal	0.87		

* Depreciation is shown as an average over the life of the plant (20 years)

7 Conclusion: Ethanol Production Today

Ethanol production technology is well developed. Conventional processes are highly outmoded and several superior alternatives are ready for pilot plant testing and industrial use. Distillation need not be a large energy consumer. The vapor reuse methods developed during WW II consume far less energy than is contained in the product alcohol and recently developed techniques may allow further savings. The major cost in alcohol production still remains the sugar source, and the development of cheap sugar sources should allow substantial alcohol price reductions.

8 Acknowledgement

The assistance of Mr. Harry Wong (Department of Chemical Engineering, University of California, Berkeley) in providing information on yeast growth and nutritional requirements is gratefully acknowledged.

9 References

1. *Chem. Eng. News 51*, No. 10, 7 (1973)
2. Hough, J. S. et al.: Continuous Culture in Brewing. Chapter 17, 1900
3. Jacobs, P. B.: Industrial Alcohol, U.S.D.A., Misc. Publication 695, Feb. 1950
4. Rose, D.: Yeast for Molasses Alcohol, Process Biochemistry 11, (3), 10 (1976)
5. Cysewski, G. R., Wilke, C. R.: Biotech. Bioeng., *20*, 1421 (1978)
6. Maiorella, B., Blanch, H. W., Wilke, C. R.: Rapid Ethanol Prod. via Fermentation. Univ. Calif., Lawrence Berkeley Lab. Rept. 10219, Nov. (1979). Presented at the AIChE 72nd National meeting, San Francisco, CA., Nov. 29, 1979
7. Katzen, R., Associates: Grain Motor Fuel Alcohol. Technical and Economic Assessment Study. U.S. Dept. of Energy Contract EJ-78-C-01-6639, June 1979
8. Hudson, J. R.: Recent Changes in Brewing Technology, Fermentation Technology Today: Proceedings of the IVth International Fermentation Symposium, p. 6290632 (1972)
9. Yand, V., Trindade, S.: Brazil's Gasohol Program. Chem. Eng. Prog., p. 11, April 1979
10. Avies, R. S.: Industrial alcohol. Kirk Othmer Encyclopedia of Chemical Technology, Kirk, R., Othmer, D. (eds.), p. 252, 1947
11. Brenner, W. et al.: Utilization of Waste Cellulose for Production of Chemical Feedstocks via Acid Hydrolysis. Clean Fuels Biomass Wastes, Symp. Paper p. 201 (1977)
12. Harris, E. et al.: Fermentation of Douglas Fir Hydrolyzate by *S. cerevisiae*, *Ind. And Eng. Chem.* 38, No. 9, p. 896, Sept. 1946
13. Ghose, T. K., Ghose, P.: J. Appl. Chem. Biotechnol. *28*, 309 (1978)
14. Wilke, C. R. et al.: Raw Materials Evaluation and Process Development Studies for Conversion of Biomass to Sugars and Ethanol. Proceedings, Second Anual Symp. on Fuels from Biomass, Dept. of Energy meeting, Troy, N.Y., Vol. 1, June 1978
15. Wilke, C. R. et al.: Enzymatic Hydrolysis of Cellulose, Theory and Applications, Rept. to the Office of Economic Cooperation and Development, 1980
16. Wang, D., Cooney, C.: Degradation of Cellulosic Biomass and Its Subsequent Utilization for the Production of Chemical Feedstocks. Ann. Rept. to Solar Energy Research Institute, Contract No. EG-77-S-02-4198 (1980)

17. Montencourt, B.: Rutgers University, Personal Communication
18. Pirt, S. J., Kurowski, W.: J. Gen. Microbiol. *63*, 357 (1970)
19. Holzer, H.: Aspects of Yeast Metabolism. Mills, A. K. (ed.), Oxford Blackwells Scientific Publications 1968
20. Akbar, M. D. et al.: Biotech. Bioeng. *16*, 455 (1974)
21. Moss, F. J. et al.: Biotech. Bioeng. *13*, 63 (1971)
22. Richard, P. D. et al.: Biotech. Bioeng. *13*, 164 (1971)
23. Rogers, P. J., Steward, P. R.: J. Gen. Microbiol. *79*, 205 (1973)
24. Haukeli, A. D., Lie, S.: J. Inst. Brew. *79*, 55 (1973)
25. David, M. H., Kirsop, B. H.: J. Inst. Brew. *79*, 20 (1973)
26. Bloomfield, D. K., Black, K. J.: Bio. Chem. 235, 337 (1960)
27. Cysewski, G. R.: Fermentation Kinetics and Process Economics for the Production of Ethanol. Ph. D. Thesis, Univ. Calif., Dept. of Chem. Eng., Berkeley, LBL-4480, March 1976
28. White, J., Munns, D. J.: Vallestein Commun., 14, 199 (1951)
29. Cowland, T. W., Maule, D. R.: J. Inst. Brew. *72*, 489 (1966)
30. Andreasen, A. A., Stier, J. J. B.: J. Cell. Comp. Phys. *41*, 23 (1953)
31. Aigar, A. S., Ludeking, R.: Chem. Eng. Progr. Symp. *62* (69), 57 (1969)
32. Bazua, C.: Effect of Alcohol Concentration on Kinetics of Ethanol Production by *Saccharomyces cerevisiae*. M.S. Thesis, Univ. of Calif., Dept. of Chem. Eng., Berkeley 1975
33. Ghose, T, K., Tyagi, R. D.: Biotech. Bioeng. *21*, 1401 (1979)
34. Aiba, S. et al.: Biotech. Bioeng. *10*, 845 (1968)
35. Holzberg, I. et al.: Biotech. Bioeng. *9*, 413 (1967)
36. Pirt, S. J.: Principles of Microbe and Cell Cultivation. p. 126, Holstead Press, Div. John Wiley & Sons, New York 1975
37. Suomalainen, H., Oura, E.: Yeast Nutrition and Solute Uptake. In: The Yeasts, Vol. 2, Chapter 2, Rose, A. H., Harrison, J. J. (eds.), Academic Press, New York 1971
38. Eddy, A. A.: Aspects of the Chemical Composition of Yeast. In: The Chemistry and Biology of Yeasts, Chapter 5, Cook, A. H. (ed.), Academic Press, New York 1958
39. Oura, E.: Biotech. Bioeng., *16*, 1197 (1974)
40. Harrison, J. S.: Yeast Production. In: Progress in Industrial Microbiology, Vol. 10, p. 129, Hockenhull, D. J. D. (ed.), 1971
41. Cysewski, G. R., Wilke, C. R.: Biotech. Bioeng. *19*, 1125 (1977)
42. Delrosario, E. et al.: Biotech. Bioeng. *21*, 1477 (1979)
43. Reuss, M. et al.: Europ. J. Appl. Microbiol. Biotechnol., *8*, 167 (1979)
44. Webster, I. A., Shuler, M. L.: Biotech. Bioeng. *21*, 1725 (1979)
45. Nagodawithana, T. W. et al.: Appl. Microbiol. *28*, 383 (1974)
46. Nagodawithana, T. W., Steinkraus, K. H.: Appl. Environ. Microbiol. *31*, 158 (1976)
47. Hough, J. S.: Production of Beer by Continuous Fermentation. Brewing Industry Research Foundation, Nutfield, Surrey, England 1959
48. Yarovenko, V. L.: Theory and Practice of Continuous Cultivation of Microorganisms in Industrial Alcoholic Processes. In: Adv. Biochem. Eng., Vol. 9, p. 1. Ghose, T. K., Fiechter, A., Blakebrough, N. (eds.), Springer Berlin 1978
49. Dawson, P. S. S.: Continuous Fermentations, Chapter 4, Ann. Reports on Fermentation Processes, Vol. 1, Perlman, D. (ed.), Academic Press, New York 1977
50. Levenspiel, O.: Chemical Reaction Engineering, John Wiley & Sons, New York 1977
51. Bishop, L.: J. Inst. Brew *76*, 172 (1970)
52. Falch, E., Gaden, E.: A Continuous, Multistage Tower Fermentor. I. Design and Performances Tests. Biotech. Bioeng., *11*, 927 (1969)
53. Ktai, A. et al.: Continuous Culture Using a Perforated Plate Column. Fermentation Technology Today, Terni, G. (ed.) (1972)
54. Kitai, A., Yamagata, Y.: Perforated Plate Column Fermentor. Process Biochem., Nov. 1970
55. Paca, J., Gregr, V.: Biotech. Bioeng., *19*, 539 (1977)
56. Prokop, A. et al.: Biotech. Bioeng. *11*, 945 (1969)
57. Fong, W. et al.: The Cost of Producing Fermentation Ethanol from Biomass. SRI International, presented at AIChE 72nd Meeting, San Francisco, CA, November 29, 1979
58. Sortland, L. D.: The Kinetics of Dense Culture Fermentations, Ph. D. Thesis, Univ. Calif., Dept. of Chem. Eng., Berkeley, May 1968

59. Elias, S.: Gasohol, New Role for Food Industry, p. 61. Food Engineering, Oct. 1979
60. Portno, A. D.: J. Inst. Brew. *73*, 43 (1967)
61. Cysewski, G. R.: Fermentation Kinetics and Process Economics for the Production of Ethanol, Ph. D. Thesis, Univ. Calif., Dept. of Chem. Eng., Berkeley, LBL-4480, March 1976
62. Ghose, T., Tyagi, R. D.: Biotech. Bioeng., *21*, 1387 (1979)
63. Wash, T., Bungay, H.: Biotech. Bioeng. *21*, 1081 (1979)
64. Kitai, A. et al.: Continuous Culture using a Perforated Plate Column. Proc. IV Int. Ferm. Symp., Kyoto, Japan, March 19, 1972
65. Hudson, J.: Recent Changes in Brewing Technology, Proc. IV IVF: Ferment Technology Today, Proc. IV Int. Ferm. Symp. p. 629 (1972)
66. Jirmann, F., Runkel, U. D.: Brauwelt, *107*, 1453 (1967)
67. Hough, J. et al.: J. Inst. Brew *68*, 478 (1962)
68. Greenshields, R., Smith, E.: Tower Fermentation Systems and Their Applications, Chemical Engineer, #249, May 1971
69. Royston, M. G.: Process Biochem. *1*, 215 (1966)
70. Klopper, W. J. et al.: Proc. of X Congress of Eur. Brew. Convention, Stockholm, 1965, p. 238. Elsevier Publ. Co., Holland 1965
71. Shore, D. T., Royston, M. G.: Chem. Eng. Lond., 218, CE 99 (May 1968)
72. Ault, R. G. et al.: J. Inst. Brew. *75*, 260 (1969)
73. Wick, E., Popper, K.: Biotech. Bioeng. *19*, 235 (1977)
74. Sitton, O. C., Gaddy, J. L.: Design and Performance of an Immobilized Cell Reactor for Ethanol Production. #41D AIChE 72nd Annual Meeting, San Francisco, CA, Nov. 25, 1979
75. Gerhardt, P., Gallup, D.: J. Bact., *85*, 919 (1963)
76. Schultz, J., Gerhardt, P.: Bact. Rev., *33*, #1 (1969)
77. Pirt, S. J.: Principles of Microbe and Cell Cultivation. Halsted Press 1975
78. Maiorella, B., Glenchur, T.: Univ. Calif., Dept. Chem. Eng., Berkeley, Notebook, RP 300 (1979)
79. Margaritis, A., Wilke, C. R.: Biotech. Bioeng. *20*, 709 (1978)
80. Margaritis, A., Wilke, C. R.: Biotech. Bioeng. *20*, 727 (1978)
81. Communication — Wilke, C. R., Univ. Calif., Dept. Chem. Eng., Berkeley 1979
82. Kan, J. K., Shuler, M. L.: An Immobilized Whole Cell Hollow Fiber Reactor or Urocanic Acid Production, AIChE Symp. Ser. 172, p. 31. Food Pharm. and Bioeng. 1976–1977
83. Breslau, B. R., Kilcullen, B. M.: Continuous Ultrafiltration. Amer. Dairy Sci. Assoc., 71 Ann. Meeting, Raleigh, N.C., June 21, 1976
84. Annon.: Romicon Hollow Fiber Ultra Filtration Membrane Cartridges, Romicon, Inc., 100 Cummings Park, Woburn, Mass. 01801 (1979)
85. Webster, I. A. et al.: Biotech. Bioeng. *21*, 1725 (1979)
86. Kan, J. K., Shuler, M. L.: Biotech. Bioeng. *20*, 217 (1978)
87. Vicroy, B.: Communication — Univ. Calif., Dept. Chem. Eng., Berkeley 1979
88. Robertson, C. R.: Enzymes, Microbes, and Chemical Engineers, Chem. Eng. Dept., Stanford Univ., presented at Univ. Calif. Berkeley, CA, Chem. Eng. Colloguiq, Jan. 1980
89. Blanch, H. W.: Communication — Univ. Calif., Dept. Chem. Eng., Berkeley 1979
90. Baker, D., Kirsop, B. H.: J. Inst. Brew *79*, 487 (1973)
91. Berdelle-Hilge, P.: U.S. Patent #3, 737, 323, June 5, 1973
92. Annon.: Chemicals by Yeast Fermentation, p. 259, Food Tech., Rev., #45 (Yeast for Food and Other Purposes) Johnson, Jeanne C. (ed.), Noyes Data Corp., Parkridge, N.J. 1977
93. Abbot, B.: Immobilized Cells, Ann. Reports on Fermentation Processes, Vol. 2, Chapter 5, Perlman, D. (ed.), Academic Press, New York 1977
94. Kierstan, M., Bucke, C.: Biotech. Bioeng. *19*, 387 (1977)
95. Divies, C.: French Patent #844, 766 (1977)
96. Chibata, I. et al.: Appl. Microbiol. *27*, 878 (1974)
97. Venkatasubramanian, K. et al.: "Enzyme Engineering", #3, Pye, L. K. and Weetal, H. (eds.), Plenum Press 1977
98. Griffith, W., Compere, A.: Dev. Ind. Microbiol. *17*, 241 (1976)
99. Kennedy, J. et al.: Nature *261*, 242, May 20, 1976
100. Pye, E. K., Humphrey, A. E.: The Biological Production of Liquid Fuels from Biomass. Univ. Penn., Interim report to U.S. Department of Energy, p. 79. Task 7, June–Aug., 1979

101. Pye, E. K., Humphrey, A. E.: Production of Liquid Fuels from Cellulosic Biomass, Proc. — 3rd Ann. Biomass Energy Systems Conf., p. 69. U.S.D.O.E., Solar Energy Res. Inst., Golden, Colo., June 5, 1980
102. Pye, E. K., Humphrey, A. E.: The Biological Production of Liquid Fuels from Biomass, Univ. Penn, Interim Report to U.S. Dept. of Energy, Feb.–May 1979
103. Craig, L.: Differential Dialysis, Science *144*, No. 3622 (1964)
104. Pye, E. K., Humphrey, A. E.: The Biological Production of Liquid Fuels from Biomass, Univ. Penn., Interim Report to U.S.D.O.E., June–Aug., 1979
105. Gregor, H.: Membrane Processes of Separation and Concentration in Biomass Harvesting, Production and Refining. Proc. 3rd Ann. Biomass Energy System Conf., p. 39, U.S.D.O.E. Solar Energy Res. Inst., Goldon, Colo., June 5, 1979
106. Gregor, H., Jefferies, T.: Ethanolic Fuels from Renewable Resources in the Solar Age, Annals of New York Academy of Science, Vol. 326, p. 273, 1979
107. Ramlingham, A., Finn, R. K.: Biotech. Bioeng., *19*, 583 (1977)
108. Maiorella, B., Wilke, C.: Biotech. Bioeng., *22* (8), 1749 (1980)
109. Zimmer, A.: Latest Developments in Energy Efficient Evaporation, Weigand Evaporators, Inc., 5585 Sterrett Pl., Columbia, Maryland 1979
110. O'Neil, D. J. et al.: Design Fabrication and Operation of a Biomass Fermentation Facility, *Tech. Prog.* report #2, p. 112, to U.S.D.O.E., Georgia Inst. of Tech., Atlanta Jan.–March 1979
111. Scheller, W., Mohr, B.: Net Energy Analysis of Ethanol Production, Omaha, NE., Dept. of Chem. Eng., Univ. Nebraska 1977
112. Sonnenblick, M.: Fusion, Sept. p18, 1979
113. Annon.: Power Alcohol, Process Engineering Company, Mannedorf, Switzerland 1979
114. Weust, R.: Handbook of Chemistry and Physics, 5th Ed, The Chemical Rubber Co., Cleveland, Ohio (1971)
115. Ofsuki, H., Williams, F.: Chem. Eng. Prog. Symp. Series, No. 6, Vol. 59, p. 55 (1953)
116. Suomalainen, H., Ronkainen, P.: Tech. Quart. Master Brewers Assoc. Am., *5*, 119 (1968)
117. Suomalainen, H. et al.: Aspects of Yeast Metabolism. A. Guinness Symp., Dublin, Blackwell Scientific Publications, Oxford 1968
118. Windholz, M.: The Merck Index, Ninth Ed., Merck and Co., Rahway, N.J. 1976
119. Webb, A., Ingraham, J.: Adv. Appl. Microbiol. *5*, 317 (1963)
120. Suomalainen, H.: Suom. Kemistilehti, 41A, p. 239, 1968
121. Ladisch, M.: Fermentable Sugars from Cellulosic Residues, Process Biochem. p. 21, Jan. 1979
122. Lecat, M.: Tables Azeotropigues, UCCLE-Bruxelles 1949
123. King, C. J.: Separation Processes. 2nd Ed., McGraw-Hill, N.Y. 1980
124. Norman, W.: Trans. of the Institution of Chemical Eng., *23*, 66 (1945)
125. Black, C. et al.: Azeotropic Distillation Results from Automatic Computer Calculations. Extractive and Azeotropic Distillation ACS Symp. No. 115, 1972
126. Black, C.: *Chem. Eng. Prog*, p. 78, Sept. 1980
127. Wentworth, T., Othmer, D.: Trans. Am. Inst. Chem. Engrs. *36*, 785 (1980)
128. Othmer, D., Wentworth, T.: Ind. Eng. Chem., *32*, 1588 (1940)
129. Wentworth, T. et al.: Trans. An. Inst. Chem. Engrs., *39*, 565 (1943)
130. Wentworth, T.: U.S. Patent 2,358,193, Sept. 1944
131. Hoffman, E.: Azeotropic and Extractive Distillation, Wiley, N.Y. 1964
132. Berg, L.: Chem. Eng. Progr., *65* (9) 53 (1969)
133. Black, C., Ditsler, D.: Dehydration of Aqueous Ethanol Mixtures by Extractive Distillation, Adv. Chemistry, *115*, 1 (1972)
134. Beebe, A. et al.: Ind. Eng. Chem., *34*, 1501 (1942)
135. Furter, W.: Extractive Distillation by Salt Effect, Adv. Chemistry Series, *115*, 35 (1972)
136. Jaques, D., Furter, W.: Prediction of Vapor Composition in Isobaric Vapor-Liquid Systems Containing Salts at Saturation. Adv. Chemistry Series, *115*, 159 (1972)
137. Meranda, D., Furter, W.: Salt Effects on Vapor-Liquid Equilibrium: Some Anomalies, *AIChE J.*, *20*, no. 1, 103 (Jan. 1978)
138. Hatch, L.: Ethvl Alcohol, Enhay Chem. Co., N.Y. 1962
139. Ellis, C.: The Chemistry of Petroleum Derivatives, Vol. 2, Reinhold, N.Y. 1937
140. Ladisch, M., Dyck, K.: Science, *205*, 898 (31, Aug. 1979)

141. G. Frankfurter, U.S. Patent 1,350,258, Dec. 20, 1918
142. Annon.: Production of Absolute Alcohol by the Haig Process, Iut'l. Sugar J., *35*, 266 (1933)
143. Myers, R.: Ethanol as a Fuel. Chem. and Eng. News, 4–9, (April 28, 1980)
144. Wilke, C. et al.: Ethanol Production and Recovery from Biomass Sugars. Lawrence Berkeley Laboratory, E & E Ann. Report 1980
145. Ricard, E., Guinot, H.: U.S. Patent 1,659,958, Feb. 21, 1928
146. Maiorella, B.: Production of Gasohol from A. Dilute Aqueous Alcohol Solution and Gasoline, U.C. Patent, Case No. 142–80, 1980
147. Hartline, F.: Science *206*, 41, (October 5, 1979)
148. Wentworth, T.: U.S. Patent 2,152,164, March 28, 1934
149. Othmer, D.: Ind. Eng. Chem., *28*, 1435 (1936)
150. Considine, D.: Chem. and Process Tech. Encyclopedia, p. 44, McGraw-Hill, N.Y. 1974
151. Andronikashili, T. et al.: Preparation of Absolute Ethyl Alcohol by Zedite, Zavod. Lab., Vol. 32, no. 10 1211, 1966
152. Milton, D. et al.: Gasohol Economic Feasibility Study Energy Research and Development Center, Univ. of Nebr., July 1978
153. Stone Webster Engineering Corp., Preliminary Economic Evaluation of Nebraska Grain Alcohol Plant, Agricultural Products Industrial Utilization Committee, State of Nebraska Dec. 1976
154. Lipinski, E. et al.: Systems Study of Fuels from Sugar Cane, Sweet Sorghum and Sugar Beets. Battelle, Columbus Labs, for Energy Res. and Develop. Admin., September 1976
155. Humphrey, A. et al.: Biological Production of Liquid Fuels and Chemical Feedstocks, Univ. Penn. and General Electric Co., for Congress of the United States Office of Technol. Assessment 1978
156. Harden, A.: Alcoholic Fermentation, p. 168, 1932
157. Stevens, R.: J. Inst. Brew, *66*, 453 (1960)
158. Webb, A., Ingraham, J.: Adv. Appl. Microbiol. *5*, 317 (1973)
159. Rainbow, C.: Brewers Yeast. In: The Yeasts, Vol. 13, Rose and Harrison (eds.), Academic Press, New York 1970
160. Harrison, J., Graham, J.: Yeasts in Distillery Practice. In: The Yeasts, Vol. 13, Rose and Harrison (eds.), Academic Press, New York 1970
161. Robinson, P.: Methyl Alcohol. Chem. and Process Tech. Encyclopedia. Considine, D. (ed.), McGraw-Hill, N.Y. 1974
162. Sciamanna, A. et al.: Composition and Utilization of Cellulose for Chemicals from Agricultural Residues, LBL, Report. 5966, UC-4, Dec. 1977
163. Wilke, C. et al.: Raw Materials Evaluation and Process Development Studies for Conversion of Cellulose to Sugar and Ethanol, Lawrence Berkeley Laboratory, LBL-7847, 1978
164. Falkehay, S.: J. Appl. Polymer Sci. (Appl. Polymer Symp. 28) p. 247 (1975)
165. Annon.: Lignin Conversion Process Shows Promise, Chem. Eng. News, Nov. 3, 1980
166. Stephen, D. et al.: Chemical Feedstocks and Fuels from Lignin, AIChE Symp. Series No. 181 Vol. 74, 1978
167. Fong, W. et al.: The Cost of Producing Fermentation Ethanol from Biomass. SRI Int'l, Presented at AIChE 72nd Meeting, San Francisco, CA, Nov. 29, 1979
168. Solomons, G. L.: Constituents of Fermentation Culture Medium. In: Material and Methods in Fermentation, Chapt. 3, Academic Press, New York 1969
169. Paul, B. P., Manmohar: Recovery of Waste Heat in Power Alcohol Distillation Plant. Ethyl Alcohol Production Technique, Noyes Development Corp., 1964
170. Paul, B. P., Mitra, A. K.: Efficient Use of Waste Heat in Rectified Spirit Distillation Plant. Ethyl Alcohol Production Technique, Noyes Development Corp. 1964
171. Bitford, H. R. et al.: Ind. Eng. Chem. *34* (11), 1406 (1942)
172. Wong, H. et al.: Media Development and Kinetics of Ethanol Fermentation, presented at the 79th Ann. Meeting of the American Society for Microbiology, Los Angeles, Ca., May 4–8, 1979

Conversion of Hemicellulose Carbohydrates

Cheng-Shung Gong, Li Fu Chen, Michael C. Flickinger*, George T. Tsao
Laboratory of Renewable Resources Engineering Purdue University
West Lafayette, Indiana 47907 U.S.A.

* Frederick Cancer Research Center Frederick, Maryland 21701, U.S.A.

Hemicellulose can be converted to a variety of useful products. There are two approaches to hemicellulose bioconversion; hemicellulose can be directly converted, or the hemicellulose-derived carbohydrates can be used as the substrate. The major problem in the bioconversion of hemicellulose carbohydrates is that suitable organisms which convert pentoses efficiently have not been developed. The advantage of bacterial processes is that a diverse range of products can be formed. The advantages of the yeast process is that specific products such as ethanol and polyols can be produced in high yields. The understanding of metabolic pathways and metabolic regulation is important for the improvement of existing microbial strains or the development of new strains. The use of yeasts to produce ethanol from D-xylose through isomerization of D-xylose to D-xylulose and the prospects for future developments in biomass conversion are discussed.

1 Introduction

Hemicellulose, one of the major constituents of plant materials, comprises up to 40% of all biomass. There are many potential uses for hemicellulose and hemicellulose-derived carbohydrates. Hemicellulose can be converted by microorganisms to various products, such as methane, organic acids, sugar alcohols, solvents, animal feed, and ethanol. Bioconversion of hemicellulose often requires prior hydrolysis of the polysaccharides to their sugar constituents. The yield, rate of hydrolysis, and type of sugar recovered depend on the source of substrate and its composition. The types and amounts of conversion products obtained from hemicellulose-derived sugars dependense on the organism, substrate, and culture conditions used.

Of the many products available from hemicellulose-derived carbohydrates, ethanol has recently received the most attention. This recent interest in ethanol production focuses on its potential use for blending with petroleum to make "gasohol". In addition to its use as a fuel or petroleum supplement, ethanol is also a versatile chemical feedstock, and many chemical products are derived from ethanol.

Ethanol is produced commercially by both chemical and microbial syntheses. While virtually all industrial alcohol is currently manufactured synthetically from petroleum and natural gas, all beverage alcohol is produced from grain, molasses, and other materials containing starch and sugar. The major sources for industrial ethanol production are carbohydrates in the form of grains, crops, crop residues, cellulosic materials, and industrial wastes.

Cellulose and hemicellulose are hydrolyzed to their sugar constituents by acids and microbial enzymes. The hydrolysis product of cellulosic materials contains a mixture of sugars with glucose and xylose as the major components. Many bacteria are able to assimilate and convert pentoses to a variety of products; yeasts on the other hand, utilize are not able to degrade pentoses even though they hexoses readily to produce ethanol in high yields. In order to convert biomass-derived sugars to ethanol efficiently, the conversion of both hexoses and pentoses is necessary.

For the conversion of biomass-derived sugars to products, biological conversion is preferred over chemical conversion. The advantage of biological conversion of biomass is that specific products are made in high yields at moderate temperature and atmospheric pressure. A number of biological processes have been investigated for the conversion of cellulose, starch, and sugars to fuels and chemicals while little progress has been made toward the conversion of hemicellulose-derived pentoses.

The efficiency of ethanol production from renewable biomass depends on the ability of microorganisms to fully convert the available carbon source into products. This means that as little as possible of the carbon and energy sources should be diverted into growth and cell maintenance.

Many research groups have been involved in the selection of better microbial strains for industrial ethanol production. Particularly noteworthy is the development of yeast strains that tolerate high substrate and ethanol concentrations giving high final ethanol yields. Thermophiles and high temperature-tolerant strains could also be useful, particularly when temperatures are high, since cooling problems are simplified. Furthermore, high temperature facilitates the hydrolysis of biomass and results in higher rates of ethanol production. Selection of high ethanol-tolerant flocculating strains is also important because these could be used in a continuous process.

For the direct conversion of cellulosic polysaccharides, a biological system that would hydrolyze both cellulose and hemicellulose to its constituent hexoses and pentoses, and convert these simple sugars to a single product, such as ethanol, would be ideal. Such a process would by-pass the expensive pretreatments and chemical and enzymatic hydrolyses of cellulose and hemicelluloses.

Many prokaryotes and eukaryotes produce extracellular cellulases and hemicellulases when grown in a medium containing cellulose or cellulase inducers, and they hydrolyze cellulosic materials into sugars for carbon and energy sources. However, most of them are not able to convert these sugars to ethanol. Still other organisms are able to produce ethanol efficiently from simple sugars, yet for the most part, they do not have the genetic make-up to produce hydrolytic enzymes for cellulose and hemicellulose degradation.

Ethanol has been produced from cellulose by a coupled saccharification-fermentation process that uses cellulose, cellulase, yeasts, and nutrients [1-3]. Separate steps for cellulase production and concentration are required. Mixed cultures of a cellulolytic mycelial mold and a glucose-degrading yeast or bacterium have also been evaluated for the production of ethanol and microbial protein from cellulose [4-6]. Similarly, a mixed culture of a cellulolytic bacterium, *Thermoactinomyces*, and ethanol-producing bacterium, *Clostridium thermocellum*, has produced ethanol from cellulosics in a single saccharification-fermentation step [7].

For the direct conversion of cellulosic materials to ethanol, a thermophilic anaerobic bacterium, *C. thermocellum*, LQ8 has been used in combination with a thermophilic pentose-degrading anaerobe. *C. thermosaccharolyticum*, to hydrolyze and convert cellulosic materials to ethanol. This mixed culture has been shown to transform both Solka-Floc and corn stover to a mixture of fermentation products that contained ethanol, acetic and lactic acids [8,9]. Recently, a mycelial mold, *Monilia sp.* has been reported to transform cellulose and hemicellulose to ethanol [10]. However, the rate of direct conversion of cellulosic materials is slow due to the polymeric nature of cellulose and hemicellulose.

Since D-xylose is the basic backbone and is the major sugar constituent of plant hemicellulose, this review will focus on the metabolism and conversion of D-xylose. The discussion will also stress on ethanol production by yeasts, especially in light of the recent success with yeast fermentation of D-xylose-derived D-xylulose to ethanol [11,12]. The detailed information related to bacterial conversion of pentoses has recently been reviewed [13-16].

2 Hemicellulose

Hemicellulose is often described as plant cell-wall polysaccharides that are associated with cellulose in lignified tissues [17]. The close association of hemicellulose with cellulose and lignins contributes to cell-wall rigidity and flexibility. The majority of the hemicellulose polysaccharides are derived from cell-wall middle lamella. Some of the non-starch, non-cellulose polysaccharides, excluding pectic materials which are known as cereal and pentosans, are sometimes also considered hemicellulose [18]. Hemicelluloses are composed of various hexoses, pentoses, uronic acids and other minor sugars. Thus hemicelluloses by definition are the short branched-chain heteropolysaccharides of mixed hexosans and pentosans that are easily hydrolyzed. D-Xylose and L-arabinose are the major constituents of pentosans while D-glucose, D-mannose and D-galactose are the constituents of hexosans [19].

Hemicellulose may be separated into two basic fractions, hemicellulose A and B, based on solubility in alkaline solutions [20]. However, there are no other clear distinctions between the two types except that hemicellulose B usually contains a higher proportion of uronic acid than hemicellulose A.

2.1 Structure and Composition

Unlike cellulose, starch or pectic materials, hemicellulose shows variability in both structure and constitution. The degree of polymerization of short-chained heteropolymers of hemicellulose is usually less than 200 [17]. Thus hemicellulose chains are simple or mixed polysaccharides of smaller dimension than cellulose.

The interior chain of hemicellulose consists of polysaccharides that are attached to a variety of sugar residues that are the same or different from the sugars that form the side chains. Most hemicelluloses contain two to six different sugar residues. The types of hemicelluloses are often classified according to the sugar residues present [17, 21]. Commonly occurring hemicelluloses are D-xylan, L-arabino-D-xylan, L-arabino-D-galactan, L-arabino-D-glucurono-D-xylan, L-O-methyl-D-glucurono-D-xylan, L-arabino-(4-O-methyl-D-glucurono)-D-xylan, D-gluco-D-mannan, and D-galacto-D-gluco-D-mannan.

L-Arabinans are often associated with the pectic materials but usually are considered to be hemicellulose. Except for the galactose-based hemicelluloses, which are β-1.3-linked, the structure of most hemicellulose is β-1-4-linked [17]. The detailed structure and composition of hemicellulose have been reviewed [17, 19, 21–23].

The type and amount of hemicellulose varies widely, depending on plant materials, type of tissue, stage of growth, growth environment, physiological conditions, storage, and method of extraction [24–26]. For these reasons, it is difficult to obtain a typical sugar composition of a typical hemicellulose. By far the most abundant type of hemicellulose has a D-xylose backbone with L-arabinose as side-chain. Table 1 summarizes the amounts of hemicellulose in different plant materials and their derived products.

Table 1. Biomass constituents[a]

Type of material	Hemicellulose %	Cellulose %	Lignins %
Monocotyledons			
Stems	25 ~ 50	25 ~ 40	10 ~ 30
Leaves	80 ~ 85	15 ~ 20	—
Fibers	5 ~ 20	80 ~ 95	—
Woods			
Hardwood (angiosperms)	24 ~ 40	40 ~ 55	18 ~ 25
Softwood (gymnosperms)	25 ~ 35	45 ~ 50	25 ~ 35
Papers			
Newspaper	25 ~ 40	40 ~ 55	18 ~ 30
Wastepaper	10 ~ 20	60 ~ 70	5 ~ 10
Waste fibers	20 ~ 30	60 ~ 80	2 ~ 10

[a] From Cowling and Kirk [29]

Table 2. Hemicellulose neutral carbohydrate content of agricultural residues[a]

Plant residues	% of total sugars			
	Xylose	Arabinose	Glucose	Others[b]
Corn residues				
Cobs	65.1	9.6	25.3	—
Leaves	59	9.4	29.1	2.5
Stalks	70.5	9.0	14.5	5.9
Husks	53.5	12.3	32.6	1.6
Pith	71.5	9.8	15.7	3
Fibers	63.8	6.6	26.8	2.8
Wheat straw	57.9	9.1	28.1	5
Soybean				
Stalks and leaves	59.9	6.6	6.1	27.4
Hulls	26.6	12.7	21	39.7
Sunflower				
Stalks	60.6	2.2	32.6	4.6
Pith	10.7	11.8	63.5	14
Flax straw	64.6	12.8	1.2	21.4
Sweet clover hays	49.3	21.9	8.9	9.9
Peanut hulls	46.3	5	46.6	2.1
Sugar cane bagasse[c]	59.5	14.5	26	—

[a] data calculated from Krull and Inglett [28]
[b] Mannose and galactose
[c] data from Laboratory of Renewable Resources Engineering, Purdue University

The major class of hemicellulose is xylans, which are found in large quantities in annual plants and deciduous trees and in smaller quantities in conifers. Glucomannans are more abundant in conifers [19]. Xylans of grasses and cereals are generally characterized by the presence of L-arabinose linked as a single unit side-chain to a D-xylose backbone [21]. Substantial differences in sugar constituents are found in wood xylans. Wood xylans are characterized by the presence of 4-O-methyl-D-glucuronic acid linked to a D-xylose backbone. In general, the proportion of 4-O-methyl-D-glucuronic acid is higher in softwood than in hardwood [21]. Xylans have also been found in algae; the structure of xylans is green algae is basically a β-1-3-D-xylopyranan coiled helically to form microfibrils [27]. Hydrolysis of hemicelluloses in annual plants and agricultural wastes produces pentoses, primarily D-xylose, as the major products (Table 2).

2.2 Hydrolysis

Hemicellulose can be hydrolyzed to its sugar constituents by chemical or microbial processes. A wide range of microorganisms produce different types of hemicellulases in response to the different types of hemicellulose in their environments.

Because hemicelluloses are heterogeneous, with different constituents linked by different types of bonds, the enzymatic hydrolysis requires several enzymes. Each enzyme attacks one or more types of bonds and all are identified as hemicellulases. The total number of hemicellulases and the role of each enzyme are not clear. In combination, hemicellulase enzymes can hydrolyze hemicellulose to its constituent sugars. The complexity of hemicellulases and their mode of enzymatic action in various types of hemicelluloses have been recently reviewed [30].

The chemical hydrolysis of hemicelluloses is much easier to accomplish than the hydrolysis of cellulose due to the heterogeneous structure and composition of hemicellulose and its low degree of polymerization. Many acids are known to be good hydrolytic agents. The common method of acid hydrolysis uses dilute acid [31-33]. One of the earliest commercial hydrolysis processes was a dilute sulfuric acid process carried out at a relatively low temperature for a prolonged period of time. Recently, a great deal of research has examined the dilute acid hydrolysis of woods and agricultural residues to produce sugars.

During acid hydrolysis of hemicellulose, pentoses are degraded rapidly to furfural and condensation by-products [34]. In order to prevent the decomposition of sugars, especially pentoses, a more dilute acid, a shorter reaction time, a lower temperature, and the rapid removal of hydrolytic agents are required. Thus, an efficient process has been developed recently to hydrolyze hemicellulose by dilute acids at moderate temperature and atmospheric pressure [35].

3 Bacteria

Bacteria, like all other living organisms, require nutrients for growth. Essential nutrients supply bacteria with an energy source and elements for macromolecular biosynthesis. Of various forms of energy sources available, bacteria use inorganic chemicals (e.g., soil bacteria), a light source (phototrophs), and organic compounds (heterotrophs).

Facultative anaerobes grow as aerobic heterotrophs in the presence of oxygen, while they carry out fermentative metabolism in the absence of oxygen. In contrast, obligate anaerobes grow only in the absence of oxygen; the presence of oxygen is detrimental to their biological and physiological functions. Aerobic heterotrophs couple the oxidation of organic substrates to the reduction of oxygen and nitrate. This involves mitochondrial electron transport coupled with high energy production (ATP) and converts substrates into cell materials, carbon dioxide, and water. Fermentative anaerobes carry out a variety of oxidation-reduction reactions involving organic compounds, carbon dioxide, molecular hydrogen and sulfur compounds. All of these reactions yield little ATP; therefore, the amounts of cell materials derived from substrates are small. Large proportions of substrates are converted to fermentative end products.

The production of organic acids by bacteria gives the highest theoretical substrate-to-product yields; however, the efficiency of solvent production is low ($<30\%$ conversion). This is due to the toxic effect of solvents on bacteria. Some bacteria produce ethanol as the major product. However, for most bacterial processes, a variety of products are formed. Industrial-scale production of chemicals through bacterial conversion has been assessed [36,37].

3.1 Pentose Metabolism

A wide range of pentoses and pentitols are utilized by many bacteria as carbon and energy sources. In most bacteria, the direct isomerization of aldopentoses to ketoses is the first step in pentose metabolism, such that D-xylose is converted to D-xylulose, and L-arabinose is converted to L-ribulose. In some bacteria, such as *Enterobacter sp.* and *Corynebacterium sp.*, the oxido-reduction reaction of D-xylose to its corresponding sugar alcohol, xylitol, has been reported to be the initial step of D-xylose metabolism [38,39]. Reduced pyridine nucleotides required for the reduction of D-xylose are believed to be generated from the oxidation of 6-phosphogluconate by NADPH-dependent 6-phosphogluconate dehydrogenase [38]. In other bacteria the direct phosphorylation of pentose is the initial step for pentose metabolism, such that D-ribose is phosphorylated to D-ribose-5-phosphate followed by isomerization of D-ribose-5-phosphate to D-ribulose-5-phosphate [40] and D-arabinose is phosphorylated to D-arabinose-5-phosphate [41]. The metabolism of pentoses and pentitols by bacteria produces D-xylulose-5-phosphate as a key intermediate. Figure 1 summarizes the pathway for the early steps of metabolism of some common pentoses [42]. The detailed description of pentose and pentitol metabolism by bacteria has been reported [40,42–48].

The most studied organisms with respect to the early steps of metabolism of pentoses and pentitols are *Aerobacter aerogenes (Klebsiella pneumoniae)* [40,44], *Escherichia coli* [45] and *Salmonella typhimurium* [46,47]. The enzymes involved in D-xylose metabolism in these bacteria are permease, D-xylose isomerase and D-xylulose kinase. These enzymes are all inducible, produced in response to the presence of D-xylose or other pentoses [46,48]. The synthesis of these enzymes is repressed when D-glucose is present. This repression can be overcome by the presence of CAMP [49].

3.2 Anaerobic Processes

Many bacteria convert hexoses, pentoses, oligosaccharides, starch and cellulose to a variety of products including alcohols (e.g., butanol, ethanol, isopropanol, and 2,3-butanediol), organic acids (e.g., acetic, butyric, formic, and lactic acids), polyols (e.g., arabitol, glycerol, and xylitol), ketones (e.g., acetone) and gases

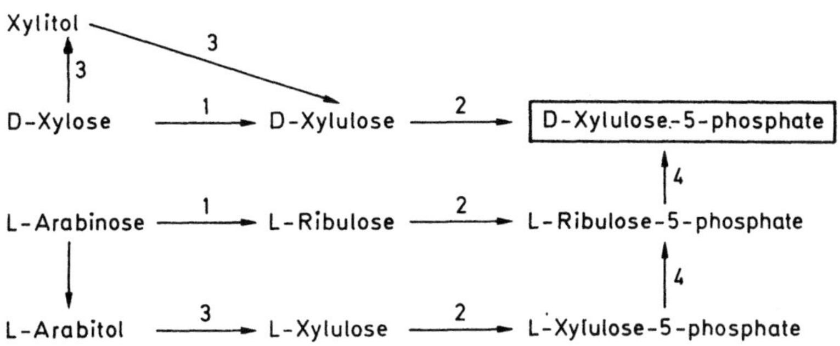

Fig. 1. D-Xylose and L-arabinose metabolism by bacteria [40]. 1, pentose isomerase; 2, pentulokinase; 3, pentitol dehydrogenase; and 4, epimerases

(e.g., CH_4, CO_2, and H_2). Anaerobic processes by bacteria result in variable rates, yields, and products, depending on the species, strain, substrate, and reaction conditions used. The diversity of metabolic pathways used for carbohydrate metabolism in bacteria makes it difficult to choose a typical organism to describe the pathway for sugar degradation; therefore, in this section we describe those pathways related to alcohol production. Several recent review articles describe organic acid production, methane production and other processes [50–56].

Unlike anaerobiosis in yeasts, which employs the Embden-Meyerhof pathway for ethanol, production metabolism in bacteria varies with respect to the pathway used. One important feature of bacteria is that most of them, produce ethanol from acetyl-CoA, the precursor of acetaldehyde [57]. Most bacteria combine the Embden-Meyerhof pathway (i.e., glycolysis) and the pentose-phosphate pathway for converting D-xylulose-5-phosphate, the key intermediate in pentose metabolism, to pyruvate. The pentose-phosphate pathway is identified by the enzymes transaldolase and transketolase. In other bacteria, (e.g., lactic acid bacteria) the ketoclastic pathway is employed in which D-xylulose-5-phosphate is cleaved to 2-carbon (acetyl phosphate) and 3-carbon (glyceraldehyde-3-phosphate) intermediates [58]. This pathway is identified by the enzyme phosphoketolase. Ethanol is derived from acetyl phosphate, and lactate is derived from glyceraldehyde-3-phosphate. The C_2 and C_3 metabolic intermediates generated by these pathways from hexoses and pentoses are transformed into fermentation end products by specific enzymes that vary with the species.

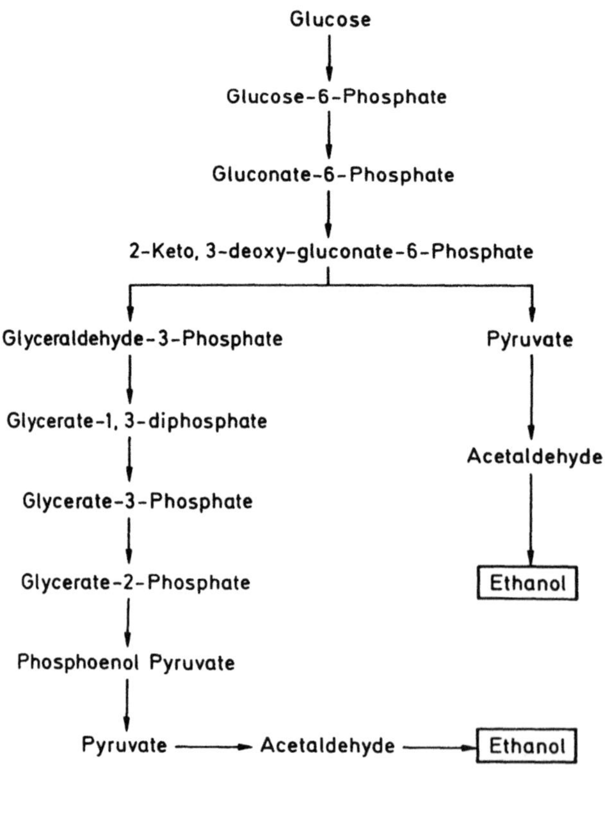

Fig. 2. Formation of ethanol from glucose by the Entner-Doudoroff pathway [59]

Bacteria that produce ethanol can be divided into several groups based upon the metabolic pathways employed.

3.2.1 Ethanol

Few bacterial species use carbohydrates to produce ethanol as the major product. *Zymomonas mobilis* and closely related species degrade glucose to pyruvate through the Entner-Doudoroff pathway [59], which is identified by the enzymes 6-phosphogluconate dehydratase and 2-keto-3-deoxy-6-phosphogluconate aldolase. The bacteria produce ethanol from acetaldehyde which is formed from pyruate by pyruate decarboxylase (Fig. 2). *Z. mobilis* strains have been reported to produce nearly 2 moles of ethanol from 1 mole of glucose [60]. None of the *Zymomonas* strains have been reported to metabolize or degrade pentoses [59]. The properties and kinetics of ethanol production by *Z. mobilis* at high glucose concentrations have recently been studied [61–64].

Sarcina ventriculi and *Erwinia amylovora* metabolize glucose through the Embden-Meyerhof pathway to produce pyruvate. Pyruvate is degraded to ethanol by the enzymes pyruvate decarboxylase and alcohol dehydrogenase [65]. *S. ventriculi* produces ethanol and small quantities of acetate, and *E. amylovora* produces ethanol and small quantities of lactate [57].

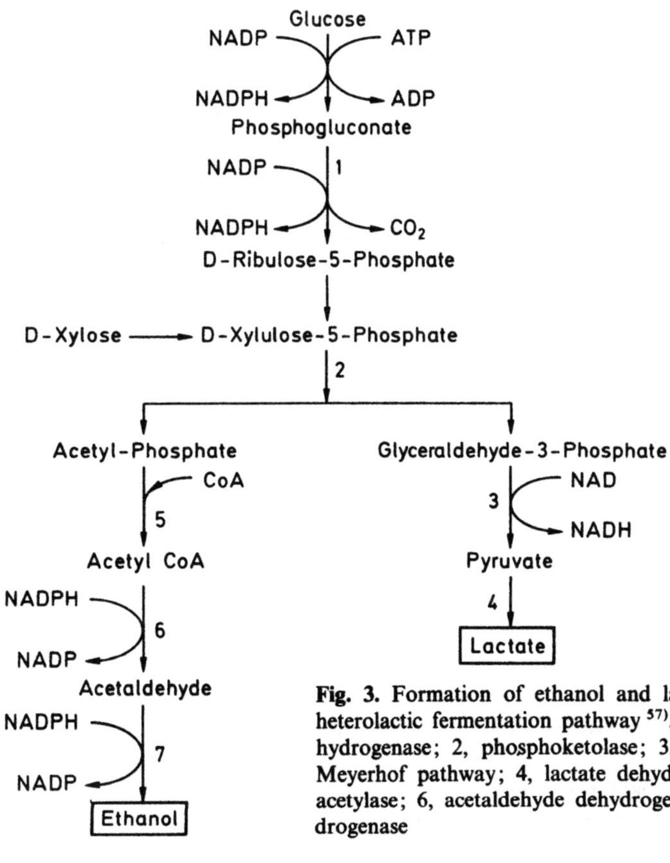

Fig. 3. Formation of ethanol and lactate from glucose by the heterolactic fermentation pathway [57]. 1, 6-phosphogluconate dehydrogenase; 2, phosphoketolase; 3, enzymes of the Embden-Meyerhof pathway; 4, lactate dehydrogenase; 5, phosphotransacetylase; 6, acetaldehyde dehydrogenase; and 7, alcohol dehydrogenase

A few anaerobic bacteria also use carbohydrates to produce ethanol as the major product [66]. A thermophilic anaerobe, *Clostridium thermohydrosulfuricum* [67], was reported to degrade glucose to ethanol quantitatively. It also uses D-xylose.

3.2.2 Ethanol and Lactate

Many bacteria, such as *Leuconostoc sp.* and *Lactobacillus sp.*, degrade sugars through the heterolactic pathway [57-68]. D-Xylulose-5-phosphate is produced from D-ribulose-5-phosphate, which is derived from phosphogluconate. D-Xylulose-5-phosphate is then cleaved into glyceraldehyde-3-phosphate and acetylphosphate biphosphoketolase (Fig. 3). Ethanol is subsequently formed from acetyl phosphate by way of acetyl-CoA and acetaldehyde, and lactate is derived from glyceraldehyde-3-phosphate by the Embden-Meyerhof pathway and lactate dehydrogenase, as in the homolactic pathway. In the course of conversion, 1 mole each of ethanol and lactate are produced from 1 mole of glucose. The product yield depends on specific growth conditions, and other products (e.g., acetate and formate) could also be formed [69-71].

Likewise, homolactic bacteria, such as *Lactobacillus casei* produce lactate, acetate, formate, and ethanol in an energy-limited continuous culture [69]. The heterolactic bacterium, *Thermoanaerobium brockii*, produces ethanol as the major product in low yeast-extract medium and produces lactate as the major product in high yeast-extract medium [72].

3.2.3 Ethanol and Mixed Acids

This type of reaction is the most common in anaerobic metabolism of carbohydrates by bacteria. The major products by this type of reaction are ethanol, acetic, lactic, succinic, and formic acids. The relative amounts of products formed depend on the organism, substrate and process conditions. Generally, bacteria are able to utilize many pentoses, pentitols, and polysaccharides and produce a variety of products. Thus, for the purpose of producing some specific products many mutant strains have been selected.

Most bacteria in mixed acid production use the Embden-Meyerhof pathway for converting pyruvate to anaerobic products. The enzymes involved in acid production are acetate kinase, pyruvate dehydrogenase, lactate dehydrogenase, acetaldehyde dehydrogenase, and ethanol dehydrogenase. The function of the regulatory properties of these enzymes depends on the specific species and the specific growth conditions. Considerable variation in pathway and end-product characterize this type of bacterial processes. Detailed studies on the regulation of mixed acid formation have been reported [71-76]. Figure 4a summarized the pathways for the mixed-acid production.

3.2.4 Butanediol and Ethanol

This type of reaction is carried out by many bacteria that are enterobacteria. The products include butanediol, ethanol, and organic acids [51, 77-82]. The bacteria combine the pentose phosphate and the Embden-Meyerhof pathways to convert hexoses and pentoses to pyruvate. All the products are derived from pyruvate (see Fig. 4b).

In addition to enteric bacteria, members of the genera *Bacillus* and *Clostridium* also produce ethanol and butanediol in various quantities [81]. The relative amounts

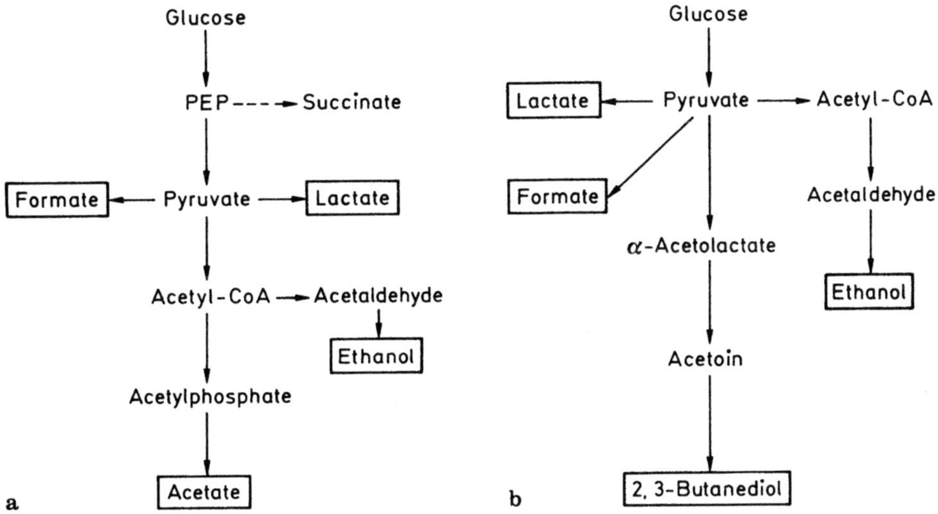

Fig. 4. Mixed acid (**a**) and butanediol process (**b**) [57]

of products formed depend on the organism, substrate composition, and process conditions. D-Xylose and L-arabinose are good substrates for butanediol production [14]. Species of genera *Aeromonas*, *Bacillus*, and *Klebsiella* produce less acid and more ethanol and butanediol [57, 80, 82]. The pH profoundly influences the products formed; at low pH the production of butanediol and ethanol is favored while at high pH the formation of organic acids is favored [81].

Quantitative assessments of ethanol and butanediol production from pentoses by many bacteria have been reported [14, 80, 82]. Processes for the production of ethanol and butanediol from pentoses derived from biomass and waste materials have also been described [83, 84].

3.2.5 Acetone-Butanol-Ethanol

Many anaerobic bacteria degrade hexoses, pentoses, and starch to a mixture of products including acetone, butanol, and ethanol [85–87]. The general metabolic pathway is outlined in Fig. 5. Acetoacetyl-CoA is the precursor for butanol, and acetone. Acetyl-CoA is the precursor of ethanol. The type and yields of products vary depending on the type and concentration of substrate, reaction conditions, specific strain, and growth phase of bacteria [87, 88].

As in the butanediol and mixed-acid production, pH affects the type of products formed. Under low pH conditions the production of solvents increases, while under high pH conditions organic acid production is enhanced [87]. Similarly, the product formed is affected by the growth phase of the bacteria. During exponential growth, acid production predominates while during the stationary phase, solvent production is preferred [89, 90].

The solvent yields by *C. acetobutylicum* from two different substrates, glucose and xylose, are similar. The degradation of sugars by *C. acetobutylicum* produce butanol, acetone, and ethanol in a ratio of 6:3:1, respectively [85].

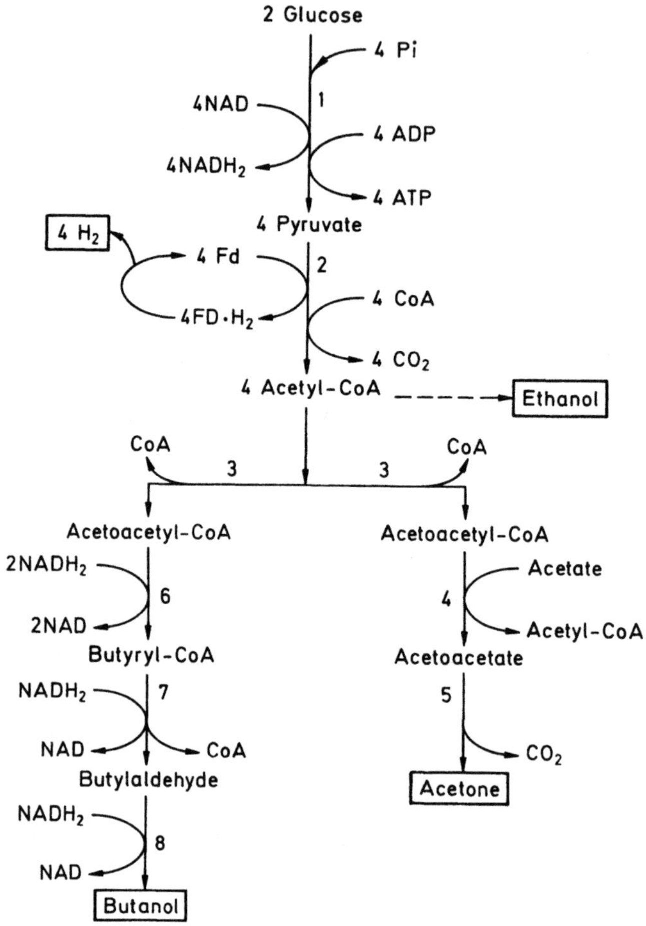

Fig. 5. Butanol and acetone production [57]. 1, enzymes of Embden-Meyerhof pathway; 2, pyruvate-ferredoxin oxidoreductase; 3, acetyl-CoA-acetyltransferase; 4, acetyl-CoA transferase; 5, acetoacetate decarboxylase; 6, L(+)-β-hydroxy-butyryl-CoA dehydrogenase, crotonase and butyryl-CoA dehydrogenase; 7, butyraldehyde dehydrogenase; and 8, butanol dehydrogenase

Many biomass-derived carbohydrates, such as those from corn cobs [91], woods [92], and waste sulfite liquors [93], have been used as substrates for this type of reaction.

3.2.6 Acetone-Ethanol

B. macerans degrade D-xylose, L-arabinose, and many hexoses to a mixture of ethanol, acetone and acetic acid [94, 95]. Ethanol and acetate are derived from acetyl-CoA, and acetone is derived from acetoacetyl-CoA (Fig. 5). A low pH favors the production of acetone and ethanol while a high pH favors acetate formation [96].

4 Yeasts

Yeasts have been used for making alcoholic beverages and bread for centuries. By definition, yeasts are "unicellular uninucleate fungi that can reproduce by budding, fission, or both and that bear a yeast-like macroscopic and microscopic appearance".

Yeasts can be further defined as "microorganisms in which the unicellular form is conspicuous and which belong to the fungi"[97]. However, since not as, yeasts are natural groups of organisms, the definitions tend to be artificial and arbitrary. Nevertheless, yeasts can be distinguished from bacteria and molds by their size, shape, mode of reproduction, and cellular and cultivation characteristics.

Yeasts are facultative organisms which have the ability to produce energy for their own use from suitable organic compounds under both aerobic and anaerobic conditions. Under aerobic conditions, sugars are metabolized to carbon dioxide and water for the production of energy and cell constituents. Under anaerobic conditions, most sugars are metabolized ultimately to ethanol by a process known as "alcoholic fermentation". The alcoholic reaction differs slightly from "glycolysis" in which glucose is degraded to 2 moles of lactic acid in the glycolysis process. The sequence of reactions of glycolysis and alcohol formation known today was determined by the pioneers in "enzymology". The biological, biochemical, and physiological aspects of yeasts in alcohol formation and sugar metabolism have been described in countless research and review articles.

The use of yeasts to convert carbohydrates derived from starchy materials, cellulosic materials, and agricultural products to ethanol has been studied extensively. Yeasts have been used in a coupled saccharification-fermentation process to produce ethanol from hexoses derived from cellulose[1-3]. Yeasts have also been used in combination with other cellulolytic microorganisms to convert cellulosic materials to ethanol. In these processes, the hemicellulose-derived pentoses are not utilized because yeasts are unable to use pentoses anaerobically. Thus, the leftover pentoses are aerobically converted to single-cell-protein by pentose-utilizing yeasts such as *Torula* and *Candida*[5,33]. In order to maximize ethanol production from biomass, the use of a single organism, such as yeast, to convert both hexoses and pentoses to ethanol is preferred.

4.1 Pentose Metabolism

Procaryotic and eucaryotic microorganisms have different mechanisms of pentose assimilation. Most bacteria isomerize the aldopentose to its ketoisomer, whereas yeasts and mycelial molds metabolize the pentose through an oxidation-reduction reaction[98-100]. In yeasts, D-xylose is reduced to xylitol by $NADPH_2$-dependent aldoreductase, and xylitol is then oxidized to D-xylulose by NADP-dependent D-xylulose reductase[101]. D-Xylulose is then phosphorylated to D-xylulose-5-phosphate, which is finally converted to pyruvate through both the pentose-phosphate and Embden-Meyerhof pathways. The pentose-phosphate pathway is identified by transaldolase and transketolase enzymes[102]. The initial $NADPH_2$ required for the reduction of D-xylose to xylitol is probably provided by the metabolism of stored carbohydrates. $NADPH_2$ is generated by the oxidation of glucose-6-phosphate to D-ribulose-5-phosphate in the pentose-phosphate pathway (Fig. 6). Since D-fructose-6-phosphate is in rapid equilibrium with D-glucose-6-phosphate, the oxido-reduction reaction of D-xylose to D-xylulose will ultimately provide $NADPH_2$ from its own reaction[99].

Yeasts also metabolize L-arabinose by the oxido-reduction reaction in which

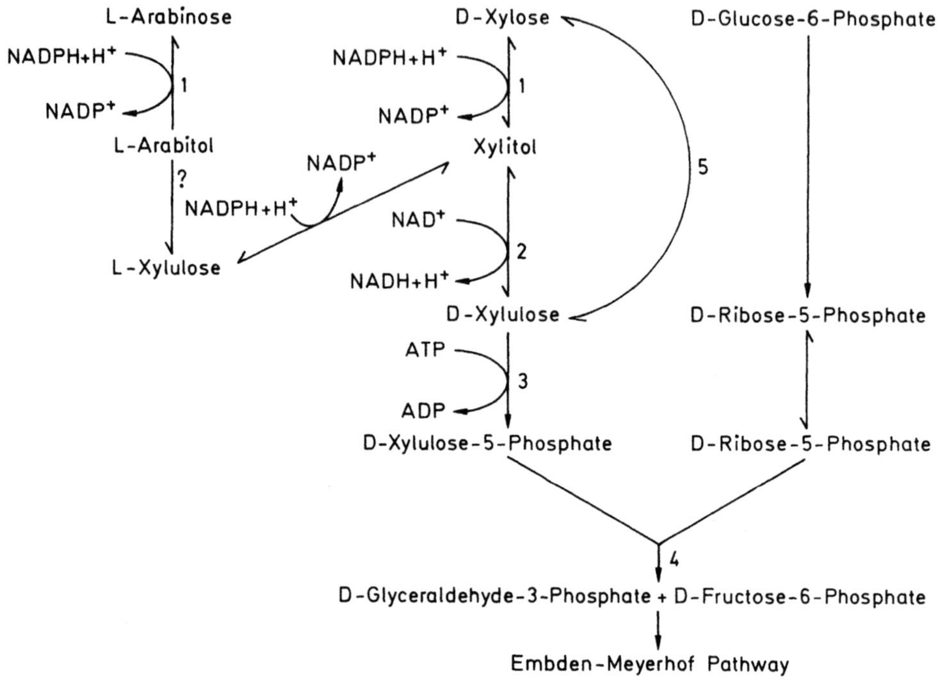

Fig. 6. D-Xylose and L-arabinose metabolism by yeasts. 1, aldoreductase; 2, D-xylulose reductase; 3, D-xylulose kinase; 4, transaldolase and transketolase; and 5, D-xylose isomerase

probably the same enzyme, aldoreductase [103], reduces L-arabinose to L-arabitol, which is converted to the key intermediate of the pentose-phosphate pathway, D-xylulose-5-phosphate (Fig. 6).

In D-xylose metabolism, sugar is taken up and accumulated without being metabolized. The actual metabolism of D-xylose occurs after a lag period of adaptive processes presumably for the production of D-xylose metabolizing enzymes [101]. The uptake of D-xylose, D-glucose, D-galactose, and D-arabinose is directed by the same carrier while the uptake of D-fructose employs a different carrier [104]. Two D-xylose carrier systems with different kinetic constants have also been reported to be present. The lower K_m carrier is repressed in rapidly growing cells and becomes derepressed by starvation [105].

In some yeasts, the presence of the inducible enzyme, D-xylose isomerase, has been demonstrated [101, 106]. The direct isomerization of D-xylose to D-xylulose as the initial step of D-xylose metabolism could occur in addition to the oxido-reduction reaction in certain yeasts.

Recently, Gong et al. [12, 108] and Wang et al. [11, 107] reported that many yeasts use D-xylulose (the keto-isomer of D-xylose) to form ethanol in high yields. The degradation of D-xylulose to ethanol by yeasts has many properties in common with that of glucose [12]. The investigators also showed that D-xylulose can be produced from D-xylose in the presence of the commercial enzyme, D-glucose isomerase [12, 109]. These results indicate that the pathways for D-xylulose metabolism are functional in yeasts.

Since the presence of the inducible enzyme, D-xylose isomerase, has been demonstrated in some yeasts, the direct isomerization of D-xylose to D-xylulose may occur in certain yeasts. However, the oxido-reduction reaction seems to be an obligatory step in the pentose metabolism of yeasts [100]. The general inability of yeasts to use pentoses may be due to a metabolic regulation mechanism that prevents anaerobic utilization of pentoses. A similar mechanism, the "Kluyver Effect" has been described by Sims and Burnett [110] for oligosaccharide and D-galactose utilization by yeasts.

4.2 Conversion of D-Xylulose

Many yeasts strains examined by Gong et al. [12] degrade D-xylulose to ethanol readily (Table 3). Yeasts strains that are able to use D-glucose for ethanol formation also convert D-xylulose to ethanol. In addition, similar properties were observed in D-xylulose and D-glucose degradation. The yields of ethanol from D-xylulose varied depending on the yeast strain used. Many yeast strains belonging to *S. cerevisiae* convert D-xylulose to ethanol in high yields (<85%), while yeasts belonging to species of *Candida* produce less ethanol. The reason for low ethanol yields by *Candida* yeasts is that significant quantities of xylitol are also produced from D-xylulose which compete with the ethanol production from D-xylulose [12,111]. Normally, yeasts metabolize xylose through the oxido-reduction reaction with xylitol as metabolic intermediate. The red yeasts, such as *Rhodotorula*, cannot ferment D-glucose nor D-xylulose. The inability of red yeasts to ferment sugars has been suggested to be due to their lack of the enzyme phosphofructokinase [112,113], even though the pentose-phosphate pathway has been found to be the main pathway of sugar catabolism in red yeasts [114].

Table 3. Production of ethanol and xylitol by yeasts [12]

Yeast strain (ATCC No.)	% Ethanol produced		% Xylitol produced	
	Glucose	D-Xylulose	D-Xylose	D-Xylulose
Candida diddensii (20213)	1.23	0.41	0.22	0.05
C. utilis (9256)	2.85	0.39	0.31	0.22
Rhodosporidium toruloides (26194)	0	0	0.36	0
Rhodotorula glutinis var. *salivaria* (34295)	0	0	0	0
Saccharomyces carlsbergensis (26602)	2.93	1.39	0.19	0.07
S. cerevisiae (24860)	2.85	1.08	0.17	0.08
S. cerevisiae (Baker's Yeast)	2.89	1.0	0.11	0.05
S. cerevisiae (respiratory defficient mutant)	2.96	0.77	—	0.04
S. diastaticus (28338)	2.89	0.69	0.25	0.78
Saccharomycopsis fibuligera (32693)	0.15	0	0.33	0.05
Torula candida (20214)	0.3	0.23	0.17	0.2
Trichosporon melibiosaceum (28580)	0.78	0	0.58	0.6

Table 4. Optimal conditions for D-xylulose degradation by Baker's yeast (*S. cerevisiae*) [111]

Process parameter	
Cell density	20 ~ 75 g l^{-1}
Temperature	35 ~ 37 °C
pH	4 ~ 6
Substrate (D-xylulose)	80 g l^{-1}

The process conditions influence the rate and yields of ethanol production from D-xylulose [111]. Table 4 summarizes the optimal conditions for *S. cerevisiae* to convert D-xylulose to ethanol. For D-xylulose production a temperature range of 35 to 40 °C and a pH range of 4 to 6 are optimal. Beyond this temperature and pH range, the rate and yields of ethanol production are considerably lower because of by-product formation (mainly xylitol) and yeast activity. The yeast cell density influences the rate and yields of ethanol from D-xylulose. At a lower density of yeast cells, the reaction time was significantly increased while at a higher density reaction time was decreased and ethanol yield was increased [111].

4.3 Isomerization-Fermentation of D-Xylose

Although D-xylulose is converted to ethanol by yeasts, the enzyme D-xylose isomerase must be present for the conversion of D-xylose to ethanol. Table 5 summarizes the effects of D-xylose isomerase from various microbial sources on ethanol production by yeasts. The coupled isomerization-fermentation of D-xylose by yeasts must be conducted under temperature and pH conditions that are a compromise between the optimal conditions for D-xylose isomerization and D-xylulose degradation. The optimal conditions for the coupled process are pH 6 and 37 °C [111] Separate steps of isomerization and conversion of D-xylose by isomerase and yeasts could also be employed to produce ethanol from D-xylose.

A scheme for ethanol production from D-xylose and hemicellulose hydrolysate is shown in Fig. 7. The presence of hexoses in hydrolysate do not affect the isomerization of pentoses such as D-xylose, because hexoses are degraded by yeasts to ethanol more rapidly than D-xylose is isomerized to D-xylulose [111].

Table 5. Effect of glucose isomerase on ethanol production from D-xylose by *S. cerevisiae* [12]

Organism	Isomerase	Source	% EtOH (w/v)
(Control)	—	—	0
Actinoplanes missouriensis	Cell-free	ATCC 14538	1.81
Bacillus sp.	Whole-cell	Novo Ind., Inc.	1.30
Bacillus sp.	Immobilized whole cell	Novo Ind., Inc.	1.26
Streptomyces sp.	Cell-free	ATCC 21175	1.95
S. olivaceus	Immobilized whole-cell	Miles Lab.	2.03

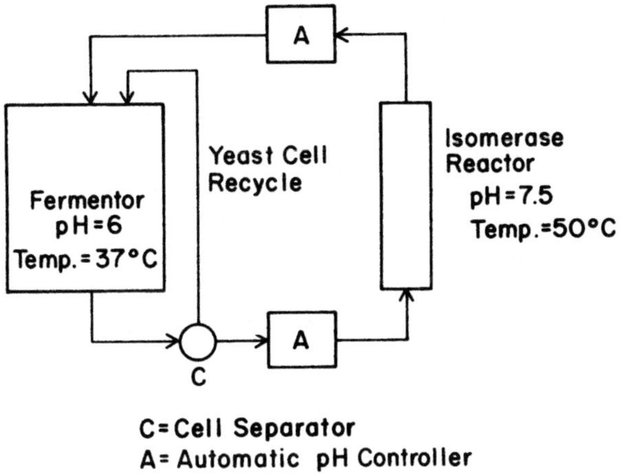

Fig. 7. Diagram of the reactor set-up for coupled isomerization-fermentation of D-xylose to ethanol by yeast and isomerase

4.3.1 D-Xylose Isomerase (Glucose Isomerase)

D-Xylose isomerase, which catalyzes the reversible isomerization reaction of D-xylose to D-xylulose was reported in the 1950's to be present in several bacteria, such as *Pseudomonas hydrophila* [115,116], *Lactobacillus pentosus* [117], and *Pasteurella pestis* [118]. Later it was found that the same enzyme also isomerized glucose to fructose and other simple sugars from their aldo to keto forms [119,120]. The synthesis of D-xylose isomerase depends on the presence of a xylose source in the growth medium. The enzyme is heat-stable, active between 45 and 78 °C and pH 6 to 8.5, and does not require any regenerating co-factors. D-Xylose isomerase is a potential enzyme source for producing the sweetener, D-fructose, from D-glucose on an industrial scale.

The presence of D-xylose isomerase is wide spread; it is produced in most prokaryotic microorganisms capable of growing on a D-xylose source. This enzyme produced from several microbial sources such as *Streptomyces*, *Bacillus*, *Actinoplanes*, and *Arthrobacter* has been studied extensively (see review by Antrim et al. [121]). D-Xylose isomerase has also been reported to be present in plant sources [122,123] and

Table 6. Commercial glucose isomerase

Enzyme source	Enzyme form	Company	Ref.
Actinoplanes missouriensis	Whole-cell	Anheuser-Busch, Inc.	124)
Arthrobacter sp.	Whole-cell	R. J. Reynolds, Co.	125)
Bacillus coagulans	Whole-cell	Novo Ind., Inc.	126)
B. coagulans	Immobilized (gelatin)	Novo Ind., Inc.	12)
Streptomyces olivaceus	Whole-cell	Miles Lab.	
S. olivochromogenes	Immobilized (alumina)	CPC Int., Inc.	127)
S. phaeochromogenes	Immobilized (polymer)	Sanyu Engineering	128)
S. sp.	Cell-free	Clinton Corn Proc., Inc.	129)
S. sp.	Immobilized (DEAE-cellulose)	Clinton Corn Proc., Inc.	129)

yeasts [101, 106]. Some of the commercially available D-xylose isomerases are listed in Table 6. Commercial application of glucose isomerase for the production of high fructose syrup has been applied extensively.

Cell-free isomerase has been immobilized by various methods including cross-linking by glutaraldehyde [130, 131], entrapment in polymers [130, 132], and adsorption by DEAE-cellulose [129, 133] and DEAE-cellulose beads [134]. Several techniques for the whole-cell immobilization processes use either direct cross-linking of activated whole cells with glutaraldehyde, or entrapment in a matrix such as cross linked collagen, carrageenan, gelatin, or cellulose fiber. For review see Hemmingsen [135]. Immobilization of the enzyme did not significantly alter the temperature and pH optimal however, thermal stability of the enzyme was improved by immobilization [136], and the pH activity profile became broader [134].

4.3.2 Isomerization

Isomerization of D-xylose to D-xylulose is achieved by either chemical isomerization at high pH or by enzymatic isomerization. Several processes have been established for the chemical isomerization of D-glucose to D-fructose. However, none of the processes have been commercialized due primarily to the degradation of keto-sugars at high pH, causing color formation and the production of off-flavors [121, 137]. The isomerization of glucose to fructose is achieved with aluminate or aluminate with borate under high pH conditions [138, 139]. Similarly, D-xylulose could be obtained from D-xylose by similar processes [140].

Various sources of D-xylose isomerase (glucose isomerase) can be used for the isomerization of D-xylose to D-xylulose. Temperature and pH influence the stability of the enzyme and the reaction rate. The optimal pH and temperature depends on the source of the enzyme. For the isomerization of glucose to fructose, and xylose to xylulose, the optimal pH is generally greater than 7.0. However, an enzyme with a lower optimum pH (= 6.5) is known to be produced by *L. brevis* [141]. On the other hand, the enzyme from *S. phaeochromogenes* has a pH optimal range of 9.0 to 9.5 [142]. The enzyme is stable over a rather wide pH range, generally at pH above 5. The range of optimal temperature for xylose isomerase ranges from 45 °C for the enzyme from *L. brevis* [141] to 90 °C for the enzyme from *A. missouriensis* [143]. Generally, the lower the temperature, the more stable the enzyme.

When glucose was isomerized at 70 °C with the enzyme from *Streptomyces*, the equilibrium of reaction was reached at 53% to 55% fructose [144]. When D-xylose was isomerized using immobilized whole-cell *Streptomyces* (Novo Indus. Inc.) at pH 6 and 60 °C, the equilibrium was reached at 30% D-xylulose and 70% D-xylose. The equilibrium of isomerization was changed by using borate in addition to the enzyme to increase yields of fructose [144, 145]. Similarly, the presence of borate in the isomerization reaction shifted the equilibrium to favor D-xylulose production from D-xylose [116, 140, 146, 147].

When a mixture of glucose and D-xylose were used as the starting materials for enzymatic isomerization, as would be the case with a cellulose or hemicellulose hydrolyzate, D-xylose was the preferred substrate [148]. The stability of isomerase during the isomerization of hemicellulose hydrolyzates varied depending upon the substrate used, pH, temperature and oxygen content. In general, half-lives for immobilized enzymes operating at 60 to 65 °C with glucose as substrate ranged from 600 to 1500 h depending on the form of the enzyme.

4.4 Polyols

Under aerobic conditions many yeasts, especially the osmophilic yeasts, possess high potential to produce polyhydric alcohols as metabolic by-products from either glucose or D-xylose [149-156]. Glycerol, erythritol, D-arabitol and mannitol are the common polyols produced by yeasts from D-glucose. Xylitol and L-arabitol are produced from D-xylose and L-arabinose, respectively, by the reduction reaction of $NADPH_2$-dependent aldoreductase. The type and amounts of polyols produced by yeasts from D-glucose are affected by many environmental factors as well as the type of yeast strain used.

At high oxygen tension, polyol production is enhanced at the expense of ethanol production [154, 157]. Similarly, limitation of nitrogen and phosphate also favor polyol production [150, 154, 158, 159]. High concentrations of salts in the media favor glycerol production by *Pichia miso* [158, 160, 161] and *S. rouxii* [162] at the expense of erythritol production. Penitol production by yeasts is observed frequently when yeasts are grown with pentoses as carbon and energy sources. Onishi and Suzuki [156] surveyed penitol production by many yeasts and found that xylitol, L-arabitol and ribitol are readily produced from D-xylose, L-arabinose and D-ribose, respectively. The yield of pentitol based on pentose consumed often reached 40%. Recently, Gong et al. [163] described mutant strains from a *Candida sp.* that produce xylitol from D-xylose almost quantitatively (<90%). The same yeast mutants also produced high quantities of xylitol from bagasse hemicellulose hydrolyzates. Xylitol was also produced from either D-xylulose or D-glucose [164, 165]. For xylitol production from glucose, a sequential process was used. D-Glucose was first converted to D-arabitol by yeasts. D-Arabitol was then converted to D-xylulose by action of the bacterium, *A. suboxydans*. D-Xylulose thus formed was then converted to xylitol by yeasts. The yields of xylitol varied, depending upon yeast strains used as well as environmental factors. High oxygen tension, low temperature and high pH conditions encourage xylitol production from D-xylose [111].

4.5 Higher Alcohols

During ethanol production, higher alcohols are formed. Higher alcohols are formed from the corresponding α-keto acids by decarboxylation and reduction. The α-keto acids are, in turn, formed either from the related amino acids by transamination (Ehrlich Mechanism) or biosynthetically from a carbon source through the pathway that synthesizes amino acids [166-171].

The carbon skeleton of higher alcohols is derived either from amino acids or pyruvate that is produced from sugars through the glycolytic pathway [164, 172, 173]. Thus, n-propanol, isobutanol, isoamyl alcohol and amyl alcohol are formed from α-ketobutyrate, α-ketoisovalerate, α-ketoisocaproate and α-keto-β-methylvalerate, respectively. The corresponding α-keto acids are, in turn, derived from threonine,

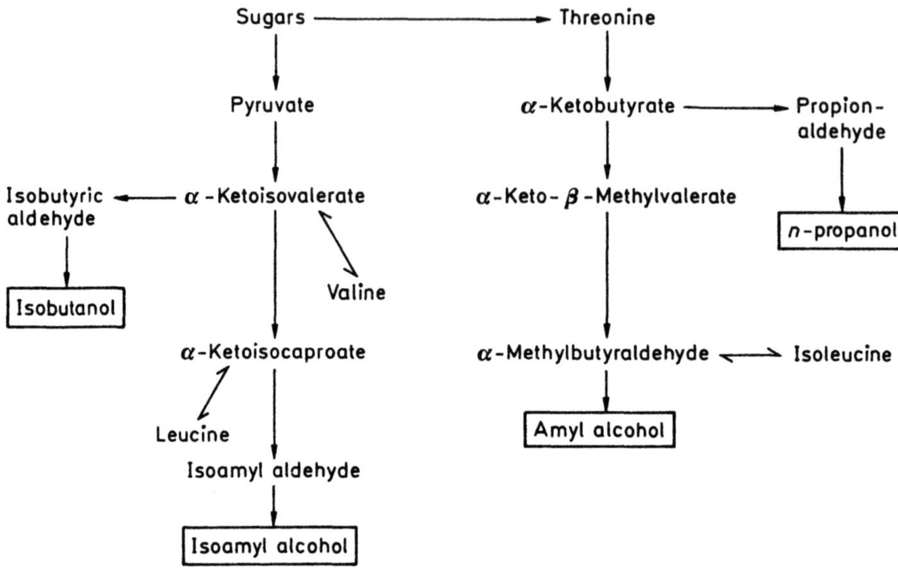

Fig. 8. Formation of higher alcohols by yeasts [169]

valine, leucine, and isoleucine, respectively. The outline of the metabolic pathway for higher alcohol formation is represented in Fig. 8.

The presence of amino acids affects the formation of corresponding alcohols through the Ehrlich Mechanism [174]. Likewise, the formation of higher alcohols is affected by many other factors, such as carbon source, nitrogen source, pH, temperature, oxygen, and yeast strains.

The regulation of higher alcohol production is complicated. The formation of isoamyl alcohol, for instance, involves a series of reactions, such as transamination of the corresponding amino acid to form the keto acid, decarboxylation of the keto acid to form the corresponding aldehyde, and then the reduction of the aldehyde to form the corresponding alcohol. The influx of amino acids, the effect of alcohols, and the incorporation of amino acids into cellular components also exert profound effects on alcohol formation [169].

5 Degradation of D-Xylose by Mycelial Fungi

Many mycelial fungi are known to metabolise hexoses and pentoses to ethanol [10, 175–177]. The early work on D-xylose metabolism by *Fusarium lini* indicated that 2-carbon and 3-carbon intermediates are formed from pentoses, which is similar to lactic acid production by bacteria (see Fig. 3). As a consequence, equal molar quantities of ethanol and acetic acid are produced [178]. The conversion of glucose by mycelial fungi is similar to that by yeast in that two moles of ethanol are produced from one mole of glucose [179].

Some members of the genus *Rhizopus* produce both ethanol and lactic acid as products [180, 181], while other members of mucoraceous fungi such as *Mucor sp.*

produce only ethanol [182, 183]. Production of lactate is possible with the presence of lactate dehydrogenase. This enzyme was first detected in the MX strain of *Rhizopus* [181] and later was partially purified from *R. oryzae* [184]. Alcohol formation is possible with the presence of alcohol dehydrogenase [181, 185].

In general, the rate of ethanol production from D-xylose by mycelial molds is much slower than that observed in bacteria and yeasts. Likewise, the rate of ethanol production from D-glucose is slow. The slow rate of ethanol production by molds could be overcome by using a high density of mycelia in a novel mycelial column [177].

A mixed culture of yeast and *Fusarium* has been used to produce ethanol from wheat hydrolyzate [186] and wood hydrolyzate [175, 187]. For ethanol production from acid-hydrolyzed wheat marshes, additional ethanol (6%—12%) is obtained from pentoses by *Fusaria* [186]. Similarly, the so-called "pentose alcohol" is produced from wood hydrolyzate by *Fusaria* in addition to the "hexose alcohol" produced by yeasts [187]. In general, *Fusaria* are very sensitive to the inhibitory material (lignins and phenolic compounds) present in the hydrolyzate [188].

Recently the use of *Fusarium* as a potential organism to degrade pentoses from hemicellulose hydrolyzates has been reexamined [189]. Additional ethanol (11%) has been obtained by using *F. oxysporum* to use pentoses after the yeast process. However, due to the slow growth rate of *Fusaria*, future developments will be necessary to improve the economic feasibility of the xylose process.

6 Conclusion

High concentrations of hemicellulose-derived carbohydrates could be easily obtained by the use of dilute acids under mild hydrolysis conditions with a low liquid-to-solid ratio. These sugars along with cellulose-derived sugars are potential substrates for fuels and chemical production.

The key to the successful conversion of hemicellulose-derived carbohydrates to fuels and chemicals lies in using the proper organisms to degrade these sugars, especially pentoses, in high yields. The conversion of pentoses by anaerobic or facultative anaerobic bacterial systems is not satisfactory at the present time due to the co-production of many products. Several mycelial fungal systems that degrade pentoses to ethanol are not practical due to their slow rates of growth and conversion.

The ideal organisms for the conversion of biomass-derived carbohydrates to products such as ethanol are the yeasts. Use of sugars by yeasts has been well studied, and many excellent industrial yeast strains have been developed and are readily available. However, yeasts are not able to degrade pentoses. The recent successful demonstration of a quantitative production of ethanol from D-xylose using yeasts and xylose isomerase would simplify the problems related to the pentose process.

The large scale production of "pentose alcohol" from agricultural residues, celluloses, and industrial wastes could be realized in the near future. Commercial yeasts, glucose isomerase (xylose isomerase) and the technology related to isomerization of sugars and alcohol production are readily available.

The pentose process technology can also be applied to grain alcohol production. For example, the combination of pentose fermentation with corn milling may allow conversion of both starch and fiber carbohydrates to ethanol in a single yeast process.

This could increase the ethanol yield up to 20%[190]. Pentose process technology could also facilitate the conversion of existing grain ethanol plants to cellulosic feed stocks. Additional alcohol can be obtained from corn cobs or corn residue, which could be transported to the ethanol plant along with the grains or could be stored for processing when grain is not available.

For the efficient conversion of hemicellulosic carbohydrates to ethanol or other products, an understanding of the metabolic pathways is important. The low yield of product is, in many cases, due to specific metabolic regulation in the specific microorganism. An understanding of metabolic regulation would help improve industrial microbial strains by improving the selection of existing strains, by enhancing product yields through mutation and selection, by manipulating metabolic pathways, and by genetic engineering.

Routine screening of new organisms as well as strain improvement programs have been successful in obtaining mutants with high production of ethanol and many other products. The selection of mutants devoid of specific enzymatic components results in a new product in the existing species or results in the accumulation of specific metabolites. This is the classical approach to mutant selection. Manipulation of metabolic pathways by controlling chemical parameters can shift the product flow and change the physiological environment of the organism to favor the production of desired products. Some examples are described below.

1) The addition of exogenous hydrogen in bacterial systems changes the electron flow and increases ethanol production[16].
2) Adding xylose isomerase to yeast systems converts nonfermentable D-xylose to D-xylulose as described in this review
3) Addition of phenylethyl alcohol to the medium stimulates ethanol production by mucoraceous fungi[191].

Through genetic engineering advances, "new organisms" can be developed so that the production of a particular metabolite is enhanced or an entirely new product is synthesized. One method of "new organism" development is by "protoplast fusion" in which a "hybrid" is created through the fusion of protoplasts from two different organisms[192,193]. A second method is through "molecular cloning" in which a foreign DNA from one of several sources is inserted into selected cells so that the integrated DNA will produce desirable products.

It is likely that much progress will be achieved in the near future in the field of biomass conversion.

Addendum

A new yeast mutant, *Candida* XF217, has recently been isolated in our laboratory. This mutant produces high yields of ethanol from D-xylose directly.

(Gong, C. S., McCracken, L. D., Tsao, G. T.: Direct Fermentation of D-Xylose to Ethanol by a Xylose-Fermentation Yeast Mutant, *Candida* sp., XF217. Biotechnol. Letters Vol. 3, No. 5, 1981).

Acknowledgement

We thank Norma Leuck, Linda McCracken and Tanya Claypool for helps.

7 References

1. Blotkamp, P. J., Takagi, M., Pemberton, M. S., Emert, G. H.: AIChE Symp. *74*, 85 (1978)
2. Takagi, M., Abe, S., Suzuki, S., Emert, G. H., Yata, N.: Proc. Bioconversion Symp. p. 551. IIT Delhi 1977
3. Meyers, S. G.: AIChE Symp. *74*, 79 (1978)
4. Han, Y. W., Dunlap, C. E., Callihan, C. S.: Food Technol. *25*, 32 (1971)
5. Peitersen, N.: Biotech. Bioeng. *17*, 1291 (1975)
6. Haggett, K. D., Choi, W. Y., Dunn, N. W.: Eur. J. Appl. Microbiol. Biotechnol. *6*, 189 (1978)
7. Pye, E. K., Humphrey, A. E.: In 3rd Annual Biomass Energy System Conference, SERI, DOE, p. 69 (1979)
8. Wang, D. I. C., Biocic, I., Fang, H. Y., Wang, S. D.: see Ref. 7, p. 61 (1979)
9. Brooks, R., Su, T. M., Brennan, M., Frick, J.: see Ref. 7, p. 75 (1979)
10. Gong, C. S., Maun, C. M., Tsao, G. T.: Biotechnol. Letters *3*, 77 (1981)
11. Wang, P. Y., Shopsis, C., Schneider, H.: Biochem. Biophys. Res. Commu *94*, 248 (1980)
12. Gong, C. S., Chen, L. F., Flickinger, M. C., Chiang, L. C., Tsao, G. T.: Appl. Environ. Microbiol. *41*, 430 (1981)
13. Doelle, W. H.: Bacteriol Metabolism, 2nd Ed., p. 738. New York: Academic Press 1975
14. Flickinger, M. C.: Biotech. Bioeng. Suppl. *1*, 27 (1980)
15. Rosenberg, S. L.: Enzyme Microb. Technol. *2*, 185 (1980)
16. Zeikus, J. G.: Ann. Rev. Microbiol. *34*, 423 (1980)
17. Whistler, R. L., Richard, E. L.: The Carbohydrates (eds., Pigman, W., Hoston, D.), Chap. 37, p. 447, New York: Acad. Press 1970
18. Aspinall, G. O.: see Ref. 17, Chap. 39, p. 515, 1970
19. Timell, T. E.: Adv. Carbohyd. Chem. *19*, 247 (1964)
20. Blake, J. D., Murphy, P. T., Richard, G. N.: Carbohydr. Res. *16*, 49 (1971)
21. Aspinall, G. O.: Adv. Carbohydr. Chem. *14*, 429 (1959)
22. Timell, T. E.: Adv. Carbohyd. Chem. *20*, 409 (1965)
23. Polyglase, W. J.: Adv. Carbohyd. Chem. *10*, 283 (1955)
24. Donnelly, B. J., Helm, J. L., Lee, H. A.: Cereal Chem. *50*, 548 (1973)
25. Wilkie, K. C. B.: Adv. Carbohyd. Chem. Biochem. *36*, 215 (1979)
26. Shafizadeh, F., McGinnis, G. D.: Adv. Carbohyd. Chem. *26*, 297 (1971)
27. Atkins, E. D. T., Preston, R. D.: Proc. Royal Soc. Ser. B *173*, 209 (1969)
28. Krull, L. H., Inglett, G. E.: J. Agric. Food Chem. *28*, 917 (1980)
29. Cowling, E. B., Kirk, T. K.: Biotech. Bioeng. Symp. *6*, 95 (1976)
30. Dekker, R. F. H., Richards, G. N.: Adv. Carbohyd. Chem. Biochem. *32*, 277 (1976)
31. Hokanson, A. E., Katzen, R.: Chem. Eng. Progress *74*, 67 (1978)
32. Porteous, A.: Paper Trade J. *156*, 30 (1972)
33. Cysewski, G. R., Wilke, C. R.: Biotech. Bioeng. *18*, 1297 (1976)
34. Wenzl, H. F.: in: The Chemical Technology of Wood, p. 157, New York: Acad. Press 1970
35. Lee, Y. Y., Lin, C. M., Johnson, T., Chambers, R. P.: Biotech. Bioeng. Symp. *8*, 75 (1979)
36. Johnson, M. J.: Ann. Rev. Microbiol. *1*, 159 (1947)
37. Owen, S. P.: Ann. Rev. Microbiol. *14*, 99 (1960)
38. Yoshitake, J., Shimamura, M., Imai, T.: Agr. Biol. Chem. *37*, 2251 (1973)
39. Yoshitake, J., Ishizaki, H., Simamura, M.: Agr. Biol. Chem. *37*, 2261 (1973)
40. Mortlock, R. P., Wood, W. A.: J. Bacteriol. *88*, 838 (1964)
41. Volk, W. A.: J. Biol. Chem. *237*, 19 (1962)
42. Mortlock, R. P., Fossitt, D. D., Wood, W. A.: Proc. Nat'l. Acad. Sci. USA *54*, 572 (1965)
43. Mortlock, R. P.: Adv. Microb. Physiol. *13*, 1 (1976)
44. Mortlock, R. P., Wood, W. A.: J. Bacteriol. *88*, 845 (1964)
45. David, J., Weismeyer, H.: Biochim. Biophys. Acta *208*, 45 (1970)
46. Shamanna, D. K., Sanderson, K. E.: J. Bacteriol. *139*, 64 (1979)
47. Shamanna, D. F., Sanderson, K. E.: J. Bacteriol. *139*, 71 (1979)
48. Wood, W. A.: Ann. Rev. Biochem. *35*, 521 (1966)
49. Pastan, I., Adhya, S.: Bacteriol. Rev. *40*, 527 (1976)
50. Lockwood, L. B.: In: Microbial Technology: Microbial Processes, (eds., Peppler, H. J., Perlman, D.), Vol. 1, Chap. 11, p. 356, New York: Acad. Press 1979

51. Pfeffer, J. T.: Biotech. Bioeng. *16*, 771 (1974)
52. Kurtzman, R. H.: Ann. Rep. Ferment. Processes *3*, 305 (1979)
53. Humphrey, A. E., Moreira, A., Armiger, W., Zabriskie, D.: Biotech. Bioeng. Symp. *7*, 45 (1977)
54. Bellamy, W. D.: Dev. Indus. Microbiol. *18*, 247 (1977)
55. Litchfield, J. H.: see Ref. 50, Chap. 4, p. 93, 1979
56. Hirose, Y., Okada, H.: see Ref. 50, Chap. 7, p. 211, 1979
57. Gottschalk, G.: Bacterial Metabolism, (ed., Starr, M. P.), p. 281, New York: Springer 1979
58. Doelle, W. H.: Bacterial Metabolism, New York: Acad. Press, 1975
59. Swings, J., Deley, J.: Bacteriol. Rev. *41*, 1 (1977)
60. Gibbs, M., DeMoss, R. D.: Arch. Biochem. Biophys. *34*, 478 (1951)
61. Rogers, P. L., Lee, K. H., Tribe, D. E.: Biotechnol. Letters *1*, 165 (1979)
62. Lee, K. J., Tribe, D. E., Rogers, P. L.: Biotechnol. Letters *1*, 421 (1979)
63. Lee, K. J., Skotnicki, M. L., Tribe, D. E., Rogers, P. L.: Biotechnol. Letters *2*, 339 (1980)
64. Cromie, S., Doelle, H. W.: Biotechnol. Letters *2*, 357 (1980)
65. Stephenson, M. P., Dawes, E. A.: J. Gen. Microbiol. *69*, 331 (1971)
66. Buchanan, R. E., Gibbons, N. E.: Bergey's Manual of Determinative Bacteriology, 8th ed., Baltimore: Williams and Wilkins 1974
67. Zeikus, J. G.: Enzyme Microbiol. Technol. *1*, 243 (1979)
68. Bryant, M. P.: In: Microbiol Energy Conversion, (eds., Schlegel, H. G., Barnea, J.), p. 107, Göttingen: Goltze 1976
69. deVries, W., Kapteijn, W. M., Van der Beek, E. G., Stouthamer, A. H.: J. Gen. Microbiol. *63*, 333 (1970)
70. Thomas, T. D., Ellwood, P. C., Longyear, V. M.: J. Bacteriol. *138*, 109 (1979)
71. Yamada, T., Carlson, J.: J. Bacteriol. *124*, 55 (1975)
72. Lamed, R. J., Zeikus, J. G.: J. Bacteriol. *141*, 1251 (1980)
73. Lamed, R. J., Zeikus, J. G.: J. Bacteriol. *144*, 569 (1980)
74. Patni, N. J., Alexander, J. K.: J. Bacteriol. *105*, 220 (1971)
75. Jungermann, K., Thauer, R. K., Leimenstoll, G., Decker, K.: Biochim. Biophys. Acta *305*, 206 (1973)
76. Wolin, M. J.: Science *146*, 775 (1974)
77. Ledingham, G. A., Neish, A. C.: Ind. Ferment. *2*, 27 (1954)
78. Wood, W. A.: In: The Bacteria, (eds., Gunsalus, I. C., Stanier, R. Y.), Vol. 2, p. 59, New York: Acad. Press 1961
79. Long, S. K., Patrick, R.: Adv. Appl. Microbiol. *5*, 135 (1963)
80. Altermatt, H. A., Simpson, F. J., Neish, A. C.: Can. J. Biochem. Physiol. *33*, 615 (1955)
81. Prescott, S. C., Dunn, C. G.: In: Industrial Microbiology, 2nd ed, p. 923, New York: McGraw-Hill 1949
82. Neish, A. C., Simpson, F. J.: Can. J. Biochem. Physiol. *32*, 147 (1954)
83. Murphy, D., Stranks, D. W.: Can. J. Technol. *29*, 413 (1951)
84. Perlman, D.: Ind. Eng. Chem. *36*, 803 (1944)
85. Spivey, M. J.: Process Biochem. *13* (11), 2 (1978)
86. Walton, M. T., Martin, J. L.: see Ref. 50, Chap. 6, p. 187, 1979
87. Peterson, W. H., Fred, E. B.: Ind. Eng. Chem. *24*, 237 (1932)
88. Compere, A. L., Griffin, D. G.: Dev. Ind. Microbiol. *20*, 509 (1979)
89. O'Brien, R. W., Morris, J. G.: J. Gen. Microbiol. *68*, 307 (1971)
90. Davies, R., Stephenson, M.: Biochem. J. *35*, 1320 (1941)
91. Langlykke, A. F., Van Lanen, J. M., Fraser, D. R.: Ind. Eng. Chem. *40*, 1716 (1948)
92. Sjolander, N. O., Langlykke, A. F., Peterson, W. H.: Ind. Eng. Chem. *30*, 1251 (1938)
93. Wiley, A. J., Johnson, M. J., McCoy, E., Peterson, W. H.: Ind. Eng. Chem. *33*, 606 (1941)
94. Northrop, J. H., Ashe, L. H., Senior, J. K.: J. Biol. Chem. *39*, 1 (1919)
95. Wilke, C. R., Blanch, H. W.: see Ref. 7, p. 79, 1979
96. Arzberger, C. F., Peterson, W. H., Fred, B. B.: J. Biol. Chem. *44*, 465 (1920)
97. Lodder, J., Kreger-van Rij, N. J. W.: In: The Yeasts, A Taxonomic Study, 2nd ed., Amsterdam and London, North-Holland 1970
98. Chiang, C., Knight, S. G.: Nature (London) *188*, 79 (1960)
99. Charkravorty, M., Veiga, L. A., Bacila, M., Horecker, B. L.: J. Biol. Chem. *237*, 1014 (1962)

100. Barnett, J. A.: In: The Fungi (eds., Ainsworth, G. C., Sussman, A. S.), Vol. III, p. 557, New York: Acad. Press 1968
101. Höfer, M., Betz, A., Kotyk, A.: Biochim. Biophys. Acta *252*, 1 (1971)
102. Osmond, C. B., Rees, T. A.: Biochim. Biophys. Acta *184*, 55 (1969)
103. Horecker, B. L.: In: Pentose Metabolism in Bacteria. New York: Wiley 1962
104. Kotyk, A., Höfer, M.: Biochim. Biophys. Acta *102*, 410 (1965)
105. Alcorn, M. E., Griffin, C. C.: Biochim, Biophys, Acta *510*, 361 (1978)
106. Tomoyeda, M., Horitsu, H.: Agr. Biol. Chem. *28*, 139 (1964)
107. Wang, P. Y., Schneider, H.: Can. J. Microbiol. *26*, 1165 (1980)
108. Ueng, P. P., Hunter, C. A., Gong, C. S., Tsao, G. T.: Biotechnol. Letters *3*, (in press) (1981)
109. Wang, P. Y., Shopsis, C., Schneider, H.: Biochem. Biophys. Res. Commu. *2*, 273 (1980)
110. Sims, A., Barnett, J. A., J. Gen. Microbiol. *106*, 277 (1978)
111. Chiang, L. C., Gong, C. S., Chen, L. F., Tsao, G. T.: Appl. Envion. Microbiol. (in press 1981)
112. Brandy, R. J., Chambliss, G. H.: Biochem. Biophys. Res. Commu. *29*, 898 (1967)
113. Höfer, M., Becker, J. U., Band, K., Deckner, K., Betz, A.: FEBS Letters *3*, 322 (1969)
114. Höfer, M., Brand, K., Deckner, K., Becker, J. U.: Biochem. J. *123*, 855 (1971)
115. Hochster, R. M., Watson, R. W.: J. Am. Chem. Soc. *75*, 3284 (1953)
116. Hochster, R. M., Watson, R. W.: Arch. Biochem. Biophys. *48*, 120 (1954)
117. Mitsuhashi, S., Lampen, J. O.: J. Am. Chem. Soc. *204*, 1011 (1953)
118. Slein, M. W.: J. Am. Chem. Soc. *77*, 1663 (1955)
119. Sanchez, S., Smiley, K. L.: Appl. Microbiol. *29*, 745 (1975)
120. Marshall, R. O., Kooi, E. R.: Science *125*, 648 (1957)
121. Antrim, R. L., Colilla, W., Schnyder, B. J.: In: Applied Biochemistry and Bioengineering: Enzyme Technology, (eds., Wingard, L. B. Jr., Katchalski-Katzir, E., Goldstein, L.), Vol. 2., p. 97, New York: Acad. Press 1979
122. Bartfay, J.: Nature (London) *185*, 924 (1960)
123. Pubols, M. H., Zahnley, J. C., Axelrod, B.: Plant Physiol. *38*, 457 (1963)
124. Gong, C. S., Chen, L. F., Tsao, G. T.: Biotech. Bioeng. *22*, 833 (1980)
125. Lee, C. K., Long, M. E.: U.S. Pat. Reissue *29*, 130 (1977)
126. Poulsen, P. B., Zittan, L.: In: Methods in Enzymology, (ed., Mosbach, K.), Vol. 44, p. 809, New York: Acad. Press 1976
127. Bengston, B. L., Lamm, W. R.: U.S. Pat. 3,654,080 (1972)
128. Kubo, T., Komatsu, T., Kawai, K., Sugihata, T., Yamada, Y.: Japanese Pat. 77,61243 (1977)
129. Lloyd, N. E., Khaleeluddin, K., Lamm, W. R.: Cereal Chem. *94*, 544 (1972)
130. Strandberg, G. W., Smiley, K. I.: Biotech. Bioeng. *14*, 509 (1972)
131. Lee, Y. Y., Fratzke, A. R., Tsao, G. T.: Biotech. Bioeng. *18*, 389 (1976)
132. Giovenco, S., Morisi, F., Pansolli, P.: FEBS Letters *36*, 57 (1973)
133. Lloyd, N. E., Khaleeludin, K.: Cereal Chem. *53*, 270 (1976)
134. Chen, L. F., Gong, C. S., Tsao, G. T.: Starch, *33*, 58 (1981)
135. Hemmingsen, S. H.: see Ref. *121*, p. 157, 1979
136. Park, Y. H., Chung, T. W., Han, M. H.: Enzyme Micro. Technol. *2*, 227 (1980)
137. Whistler, R. L., BeMiller, J. N.: Advan. Carbohyd. Chem. *13*, 289 (1958)
138. Shaw, A. J., Tsao, G. T.: Carbohyd. Res. *60*, 327 (1978)
139. Shaw, A. J., Tsao, G. T.: Carbohyd. Res. *60*, 376 (1978)
140. Hsiao, H. Y., Chen, L. F., Gong, C. S., Tsao, G. T.: (in press 1981)
141. Yamanaka, K.: Biochim. Biophys. Acta *151*, 670 (1968)
142. Tsumura, N., Sato, T.: Agric. Biol. Chem. *24*, 1123 (1965)
143. Scallet, B. L., Shieh, K., Ehrenthal, I., Slapshak, L.: Starch *26*, 405 (1974)
144. Takasaki, Y.: Agr. Biol. Chem. *35*, 1371 (1971)
145. Barker, S. A., Somer, P. J.: In: Topics in Enzyme and Fermentation Biotechnology, (ed., Wiseman, A.), Vol. 3, p. 12, New York: Wiley 1979
146. Mitsuhashi, S., Lampen, J. O.: J. Biol. Chem. *204*, 1011 (1953)
147. Campello, J. P., Veiga, L. A.: Ann. Acad. Brasil Science *45*, 309 (1973)
148. Chad, S., Ghose, T. K.: Proc. Bioconversion Symp. IIT Delhi, p. 573 (1977)
149. Hajny, G. J.: Appl. Microbiol. *12*, 87 (1964)
150. Peterson, W. H., Hendershot, W. F., Hajny, G. T.: Appl. Microbiol. *6*, 349 (1958)
151. Onishi, H.: Bull. Agr. Chem. Soc. Japan *24*, 131 (1960)

152. Onishi, H., Suzuki, T.: Appl. Microbiol. *16*, 1847 (1968)
153. Roxburgh, J. M., Spencer, J. F. T., Sallans, H. R.: Can. J. Technol. *34*, 248 (1956)
154. Spencer, J. F. T., Sallans, H. R.: Can. J. Microbiol. *2*, 72 (1956)
155. Spencer, J. F. T., Roxburgh, J. M., Sallans, H. R.: J. Agr. Food Chem. *5*, 64 (1957)
156. Onishi, H., Suzuki, T.: Agr. Biol. Chem. *30*, 1139 (1966)
157. Spencer, J. F. T., Shu, P.: Can. J. Microbiol. *3*, 559 (1957)
158. Onishi, H., Saito, N., Koshiyama, I.: Rep. Noda Inst. Japan *5*, 28 (1961)
159. Hajny, G. J., Smith, J. H., Garver, J. C.: Appl. Microbiol. *12*, 240 (1964)
160. Onishi, H.: Rep. Noda Inst. Japan *4*, 1 (1960)
161. Onishi, H.: Rep. Noda Inst. Japan *7*, 26 (1963)
162. Onishi, H.: Rep. Noda Inst. Japan *3*, 65 (1959)
163. Gong, C. S., Chen, L. F., Tsao, G. T.: Biotechnol. Letters *3*, 125 (1981)
164. Onishi, H., Suzuki, T.: Appl. Microbiol. *18*, 1031 (1969)
165. Onishi, H.: U.S. Pat. 3,619,369 (1971)
166. Ayrapaa, T.: J. Inst. Brew. *74*, 169 (1968)
167. Castor, J. G. B., Guyman, J. F.: Science *115*, 147 (1952)
168. Moutonet, M.: Ann. Technol. Agric. *18*, 249 (1969)
169. Ouchi, K., Yamamoto, Y., Takagishi, M., Akiyama, H.: J. Ferment. Technol. *58*, 301 (1980)
170. Thoukis, G.: Am. J. Enol. *9*, 161 (1959)
171. Webb, A. D., Ingraham, J. L.: In: Adv. Appl. Microbiol., Vol. 5, New York: Acad. Press 1963
172. Chen, E. C. H.: J. Am. Soc. Brew. Chem. *36*, 39 (1978)
173. Kunkee, R. E., Guyman, J. F., Crowell, E. A.: Sci. Technol. Proc. Spec. Intern. Symp. *1*, 531 (1971)
174. Engan, S.: J. Inst. Brew. *76*, 254 (1970)
175. Nord, F. F., Mull, R. P.: Adv. Enzymology *5*, 165 (1966)
176. Hesseltine, C. W.: Mycologia *57*, 149 (1965)
177. Gong, C. S., Cheng, L. J., Tsao, G. T.: Abstract, Am. Chem. Soc. 178th Ann. Meeting, September 1979
178. Gibbs, M., Cochrane, V. W., Paege, L. M., Wolin, H.: Arch. Biochem. *50*, 237 (1954)
179. Cochrane, V. M.: Mycologia *48*, 1 (1956)
180. Foster, J. W.: In: Chemical Activities of Fungi, Chap. 9, p. 648, New York: Acad. Press 1949
181. Margulies, M., Vishniac, W.: J. Bacteriol. *181*, 1 (1961)
182. Pelman, D.: Am. J. Bot. *37*, 237 (1950)
183. Gleason, F. H.: Mycologia *63*, 906 (1971)
184. Obajashi, A., Yorifuji, H., Yamagata, T., Ljichi, T., Kanie, M.: Agr. Biol. Chem. *30*, 717 (1966)
185. Stuart, T. D., LaForce, R. L., Gleason, F. H.: Mycologia *65*, 963 (1973)
186. Sciarini, L. J., Wirth, J. C.: Cereal Chem. *22*, 11 (1945)
187. Loughran, G. A., Soodak, M., Nord, F. F.: Arch. Biochem. *6*, 163 (1945)
188. Wiley, A. J., Johnson, M. J., McCoy, E., Peterson, W. H.: Ind. Eng. Chem. *33*, 606 (1941)
189. Batter, T. R., Wilke, C. R.: In: A Study of the Fermentation of Xylose to Ethanol by *Fusarium oxysporum*, Lawrence Berkeley Lab., University of Calif. 1977
190. Flickinger, M. C., Mehlberg, R. L., Gong, C. S., Chen, L. F., Tsao, G. T.: Canadian National Power Alcohol Converence, Winnipeg, Manitoba, Canada, June 1980
191. Terenzi, H. F., Stock, R.: J. Bacteriol. *97*, 1248 (1969)
192. Peberdy, J. F.: Ann. Rev. Microbiol. *33*, 21 (1979)
193. Peberdy, J. F.: Enzyme Microbiol. Technol. *2*, 23 (1980)

Fuel Ethanol from Biomass: Production, Economics and Energy

N. Kosaric, Z. Duvnjak, G. G. Stewart*
Chemical and Biochemical Engineering The University of Western Ontario
London, Ontario, Canada N6A 5B9
* Brewing Research and Development Department The Labatt Brewing Company
150 Simcoe Street, London, Ontario, Canada N6A 4M3

1 Introduction

Energy, the essential and basic commodity for human activities, has been used by man since the dawn of recorded history. While energy consumption was small and limited prior to the modern industrial era, since the industrial revolution a progressive and constant increase has been experienced (Fig. 1). A 40-fold increase of total energy usage and a 3.5-fold increase per capita in the U.S.A. between 1850 and 1976 is evident from this figure [51].

According to an EXXON report [5] a growth in energy consumption at 2.5% per year is expected in the Western world between 1980 and the year 2000. This growth rate will be less than a half of the growth rate for the period between 1965 and 1973, when it was 5.5% per year. Slower economic growth and less energy consumed per unit of output as a result of conservation, new technology and investment is expected to cause the lower rate for the future energy demand. Even with the lower consumption rate, approximately 1/3 more energy will be required in 1990 than was needed in 1978. In 2000 the world will require 2/3 more energy than in 1978.

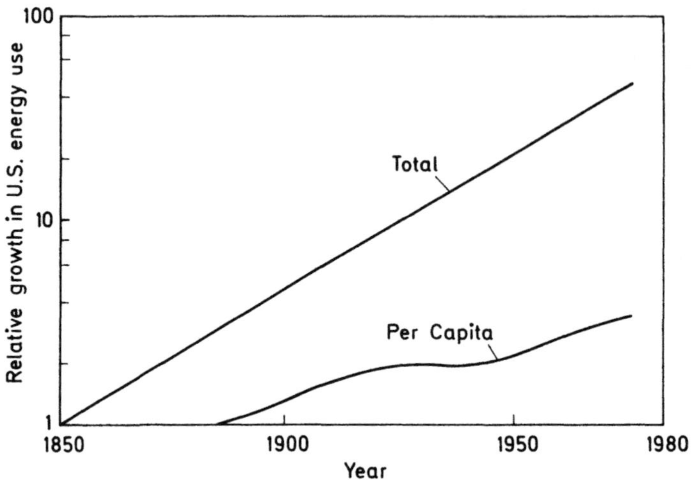

Fig. 1. Comparison of total and per capita energy use in the U.S. [51]

The general scheme of energy demand will also be changed. The U.S. demand in 1978 was 41 % of the total world demand, and it is expected that this will be dropped by about 33 % by the year 2000. Europe's share is expected to slightly decline, whilst no change in Japan's demand is anticipated. Developing countries are projected to increase their energy demand from 19 % in 1978 to 33 % in the year 2000. And developing countries with energy imports will increase their share of world energy use from 11 % in 1978 to 17 % in 2000.

The constant increase in energy consumption has already reached a level that marks the genesis of a global energy crisis. This crisis is predominantly caused by the industrially developed part of the world that today represents less than 15–20 % of the world population. It is obvious that the energy crisis would have appeared much earlier had the rest of more than 80 % of the world consumed more energy than it did.

A variety of energy forms such as coal, electricity, petroleum, natural gas, etc., were the basis of the industrial revolution. In the last 60 years crude oil was the predominant fuel and industrial raw material. The advantage related to storability, stationary and mobile use, diversity of applications, availability and the relatively low

Table 1. Energy supply for the year 2000

Energy source	% of total
Conventional oil supply	37
Natural gas	16
Synthetic fuel	4
Coal	24
Nuclear power	10
Hydro and other	9

price secured a solid place for crude oil as an industrial, agricultural and transportation fuel and as an ideal raw material for a large organic chemical synthetics industry.

Due to oil, marine and railway transportation systems were revolutionized, human and animal power were replaced in forestry and agriculture and in urban transportation; aviation and trucking transport were also developed. That development and the use of oil in industry were reasons for its intensive exploitation. The fact that oil constituted 54% of the world energy supply in 1978 demonstrates its extensive use. In spite of limited quantities of crude oil, its share in the world's energy supply will be predominant in the future [5] as predicted in Table 1.

According to EXXON estimates, the world's conventional oil resources are $1.6–2.4 \times 10^{11}$ m^3. Factors that contribute to a profound uncertainty — crude oil supply are: limiting global quantity of crude oil, unpredictable price increases and political unrest.

The only way to maintain the present industrial potential of the developed world and to allow for industrial and economic expansion of the underdeveloped world is to develop fuels that would at least partially substitute for this diminishing petroleum resource. With other fuel commodities such as coal and nuclear fuel, alcohol (methanol and ethanol) will play a role of increasing and promising potential.

Methanol and ethanol have almost all the advantages of liquid hydrocarbon fuels. However, as these fuels represent a partially oxidized hydrocarbon, their energy content per unit volume is lower than gasoline, as shown in Table 2.

Table 2. Energy content of gasoline, methanol and ethanol [31]

Fuel	Available energy
Gasoline	3.45×10^4 KJ l^{-1} (124,000 BTU/gal)
Ethanol	2.36×10^4 KJ l^{-1} (84,600 BTU/gal)
Methanol	1.80×10^4 KJ l^{-1} (64,800 BTU/gal)

The use of alcohol as a motive fuel is not a novel idea. In some European countries, the U.S.A. and in other countries around the world, between World Wars I and II there was extensive experimentation with, and use of, a variety of substitute fuels. Alcohol blends were used with success in over four million vehicles at that time, it being generally admixed in proportions of up to 25% with gasoline [37].

Industrial ethanol production was based on carbohydrates from agricultural crops. It was largely developed before World War II, especially in Europe. After 1945, cheaper ethanol from petroleum products inhibited further development of ethanol production by microbial processes and eventually replaced it. This is one of the main reasons why this technology has experienced minor improvement in the last decades. Since 1973, with the beginning of the energy crisis, more consideration has been given to the production of bioethanol, especially on the basis of renewable resources.

Due to competition with the cheap and readily available crude oil, little research in alcohol utilization for fuel was conducted prior to the October 1973 OPEC oil embargo. The embargo had tremendous consequences on the world economy and caused an uproar of dissatisfaction from the world's oil consuming nations, and brought about the realization that:

a) oil reserves are not infinite
b) low prices for this resource cannot be maintained
c) alternative fuel supplies and the associated technology have to be developed to meet increasing global energy requirements.

This last point is particularly important. If there is anyone to "thank" to this awareness, it is the OPEC's stand on crude oil availability and supply.

Several developments in fuel alcohol production have started to escalate since 1973. Brazil is a good example of this development. This country started a limited scale alcohol program before the 2nd World War and escalated it into a National Alcohol Program in 1975. The main goal of this National Program was the reduction of crude oil imports by utilizing 20% of ethanol in a blend with gasoline. Brazil's prediction is that approximately 3.4×10^9 l absolute alcohol will be required in 1980. It is also planned to produce 17,000 automobiles in Brazil in 1980 that will run entirely on absolute ethanol [12]. For comparison world alcohol production in 1975 was 5.7×10^9 l out of which 2.76×10^9 l were produced in the U.S.A. About 1.7×10^9 l or about 60% of the total production in the U.S.A. was by microbial processes. This means that the total U.S. ethanol production was about 6×10^8 l less than the predicted Brazilian consumption for 1980/81 [24, 41].

In the U.S.A. $ 8–$ 12 billion have been allocated over the next ten years for the gasohol program. The objective is to increase ethanol capacity for gasohol from 3×10^8 l at the end of 1979 to 1.2×10^9 l at the end of 1980, 1.9×10^9 l at the end of 1981 and 5.7×10^9 l by the mid 1980's. Consequently, by the end of 1981 the U.S.A. would produce 1.9×10^{10} l of gasohol. This amount represents about 10% of unleaded gasoline currently consumed in the U.S.A. [34]. However, according to the U.S. Department of Energy survey of expected new plant construction, a capacity for no more than 7.6×10^8 l of ethanol could be built this year and no more than 1.5×10^9 l in 1981 [6].

In attempting to make gasohol competitive with unleaded gasoline, the U.S. Government stimulates production of bioethanol through credits, funding, benefits, loans or loan guarantees, technical help, exemption of 4 cent federal tax, elimination of approximately 24 state gasoline taxes for gasohol, etc. The entire package of incentives gives ethanol a 13.2 cent l^{-1} subsidy. This substantially reduced the fuel's wholesale price of approx. 34.3 cent l^{-1} making gasohol competitive with the wholesale price of unleaded gasoline (22.4–23.8 cent l^{-1}).

Many other countries are also engaged in studies on ethanol production on the basis of their own natural resources. In these studies, special attention is given to "renewable materials" that are almost universally available. Prices of these raw (waste) materials are much more competitive than some of the conventional alcohol substrates such as grain. One of the most important and decisive factors in alcohol economics is the relatively excessive cost of the substrate. An advantage of renewable resources is that these are available in very large quantities, which is of critical importance for large-scale alcohol production. Availability of some of these materials is even increasing. One example is wood. According to estimates by Dever [14] 1.5×10^{10} tons more wood is available today in world forests (due to the rising amounts of CO_2 in the atmosphere) than there was a century ago.

Different vehicle tests were conducted to determine the influence of ethanol in a mixture with gasoline on fuel economy, road octane response and air pollution. It was found that ethanol in a mixture of gasoline increases the road octane value compared to gasoline. A volumetric fuel economy is slightly decreased and energy economy is increased. At temperatures up to 24 °C emission of CO, HC and NO_x is reduced, but at 38 °C it is higher when ethanol blended fuel is used [2,37]. On the basis of these results, it was concluded that ethanol is a desirable component for blending with unleaded automotive fuel.

2 Raw Materials for Ethanol Production

The following raw materials can be used for ethanol production:
sugars
starches
agricultural and forestry cellulosic materials
wastes (urban and industrial)

2.1 Sugars

Sugar cane, sugar beet and sweet sorghum can be utilized as sources of sugars for ethanol production. The first and second commodities are better known, and their juices serve for sugar manufacturing. These juices can also be used for ethanol production. However, for ethanol production on the basis of sugar cane and sugar beet, molasses is used rather than the corresponding juices. One of its advantages is that it can be stored for a longer period of time without deteriorating, which enables an off-season and possibly continuous operation of the alcohol plants.

Sweet sorghum has been grown in the U.S. for many years and used for preparation of table syrup. As demand for the syrup decreased, an interest in production of this plant also diminished. Its juice contains some starch and aconitic acid that causes difficulties in sugar crystallization. Interest in this plant as a source of degradable carbohydrates increased recently [29]. The composition of juices from sugar cane, sugar beet and sweet sorghum differ, but these differences affect sugar crystallization rather than their use as substrates for ethanol production. From an economical viewpoint, the cost of processing the plant to produce sugar juices from these three commodities is approximately the same.

An advantage of sugar juices and molasses in ethanol manufacturing is that they are ready for conversion and need no pretreatment.

2.2 Starches

All cereal grain crops, corn, potatoes, beans, field peas, cassava roots, etc., are good resources for ethanol manufacture. A high yield and starch content give corn an advantage in comparison to other commodities of this group.

Cassava, which is also known as manioc or tapioca, can be placed in this group as well. It is one of the most efficient photosynthetic plants. Cassava contains 20 to 35%

starch and 1 to 2% protein. This plant is cultivated in many tropical regions and is of interest both as a food and animal feed and as a raw material for ethanol production.

Carbohydrates in these resources are in the form of starch and have to be hydrolyzed to glucose. Chemical hydrolysis which was practiced in the past is today being widely replaced with preferred enzymatic methods.

Keeping in mind that these commodities are used for food and animal feed, moral and economic factors (except technical) have to be considered before their utilization in large quantities for ethanol production. Spoiled and low quality starches, not suitable for food or feed, can be effectively used for production of alcohol. However, even if food grade starch is being used for ethanol production, the resulting biomass byproduct would represent an excellent, protein fortified human or animal food [37].

In addition to the conventional agricultural starch resources, unicellular green algae can also be used as a substrate [47]. It has been shown that *Chlorella vulgaris*, *Chlamydomonas sp.* and *Scenedesmus basileuses* synthesize substantial quantities of starch that is contained in cells or excreted. In this cells and starches have to be pretreated before conversion to ethanol.

2.3 Wood and Residues from Agriculture and Forestry

Wood and different residues from agriculture and forestry represent an excellent resource. The net total world production of cellulose has been estimated at one hundred billion tons per year [41]. A summary of availability of organic residues and their approximate chemical composition is shown in Tables 3 and 4 [19].

Table 3. Availability of organic residues [19]

Waste/by-product	Global 1972 tons $\times 10^6$ a^{-1}	Waste/by-product	Global 1972 tons $\times 10^6$ a^{-1}
Bagasse	104	Maize	150
Barley	53	Cotton stalks	13.5
Rice straw	180	Cotton linters	1.0
Rye and oat straw	110	Wheat straw	550
Jute sticks	4.4	Wood wastes	25

Table 4. Composition of organic wastes [19]

Waste/by-product	Cellulose %	Pentosans %	Lignin %
Bagasse	45–55	25–27	19–21
Rice straw	32–53	21–24	12–25
Rice husk	35–45	25–28	15–20
Jute stick	41–58	19–22	21
Wheat straw	31–51	23–38	18–21
Hard wood	60	11	28
Soft wood	53	23	23

Large quantities of these materials make this group the richest resource for fuel ethanol production by microbial processes. Although some use has been practiced, the major portion of available quantities of these materials has not been used as a resource raw material for industry. One of the reasons is the lack of appropriate technology. Another reason is that other materials that can be processed by simpler and cheaper means were at hand. Thirdly, the demand for raw industrial materials was lower due to lower per capita and total population consumption. In the past, cellulose resources were mainly utilized for heat generation by combustion. Problems in utilization of these materials are in collecting and handling arising from their spread over large areas and from their relatively large volume per unit weight.

From their chemical composition, it is obvious that these materials can be used for ethanol production by microbial technology. However, ethanol-producing micro-organisms cannot metabolize intact cellulose. Consequently, cellulosic materials have to be pretreated. This was the main reason why these materials have not been used earlier as substrates. The presence of lignins and hemi-celluloses further complicates their utilization.

The main purposes of pretreatment processes is to convert polysaccharides of cellulosic materials to low molecular sugars that can be metabolized by micro-organisms. Both acid/alkaline and enzymatic hydrolysis processes are used for this purpose. Acid hydrolysis of wood has been practised commercially for many years. One of the earliest processes used dilute sulfuric acid and was developed by M. F. Ewen and G. H. Tomlison during World War 1; concentrated acid was also used. A process reported by Sitton et al. [39] utilized diluted followed by concentrated acid where higher yields and lower acid utilization were reported.

Enzymatic hydrolysis of cellulosic materials involves a complex of cellulases which can be obtained from fungi and bacteria [25]. In order to enhance the susceptibility of cellulose to enzymatic hydrolysis, various techniques for pretreatment of cellulosic materials have been developed. The more effective physical methods are milling, steaming and γ-irradiation, whereas sodium hydroxide and sulfur dioxide gas treatment are chemical methods employed [15,42].

2.4 Urban and Industrial Wastes

There are a number of different urban and industrial waste materials such as newspaper, food packages, cotton linters, spent sulfite liquor, cheese whey, wastes from vegetable and fruit industries, coffee wastes, etc. This waste material appears in the form of solids and liquids and has to be processed for environmental reasons. As these wastes can be used for ethanol production, their processing could even become profitable. Usage of industrial wastes for ethanol manufacturing is not a novel idea. Production of ethanol from spent sulfite liquor dates back to 1907 when the first experimental plant was built in Sweden. Modified processes have found wide acceptance, especially in Europe [18,49]. A process converting cheese whey to alcohol is also profitable and can compete even with synthetic alcohol produced from crude oil [35].

Fruit and vegetable wastes are under study for alcohol production [7,23,30,46].

Usage of urban wastes for ethanol production is also feasible due to advances in cellulase process technology, and studies have been undertaken in various research laboratories [42,44,52].

Many of the above waste materials have to be pretreated in order to yield degradable carbohydrates. Collection, transportation, concentration and storage of these materials present technological difficulties. Further, many of the industrial and agricultural waste materials are only seasonally available, which limits their continuous production potential. A positive factor is that these wastes are available in very large quantities; and their disposal represents an environmental pollution problem. When converted to a useful product like fuel alcohol, the economics for their utilization vs. disposal are most favourable.

3 Recent Modifications of Ethanol Manufacturing Processes

The best known industrial bio-process is the manufacture of ethanol. This process is based on a conversion of sugars to ethanol by different micro-organisms, but predominantly by the yeast *Saccharomyces cerevisiae*.

Conversion of sugars to ethanol can be expressed by the following biochemical equation:

$$C_6H_{12}O_6 + 2\,ADP + 2\,P \rightarrow 2\,C_2H_5OH + 2\,CO_2 + 2\,ATP\,.$$

This reaction is an exothermic process, and about 4% of the combustion energy of the sugar is released [17].

Most of the industrial bio-processes for ethanol production have been batch systems with a retention time of about 50 h. The reaction temperature is in the range of 20–30 °C and initial pH of the media is about 4.5. Ethanol yield based on sugars is about 90% of the theoretical. Concentration of ethanol in broth at the end of the process is at the level of 10–16% v/v. Sugar conversion is followed by distillation or other methods for increasing of ethanol concentration.

Several modifications of the classical alcohol processes have been studied with the aim of improved yields and higher alcohol concentrations in the broth. One of the attempts to allow a high growth and alcohol productivity is to operate the fermentor under a reduced pressure. This operation enables the alcohol to be evaporated as produced, which consequently lowers the alcohol concentration in the broth during the conversion process.

It is known that ethanol inhibits both yeast growth and sugar conversion in alcohol biosynthesis [1,16]. By operating under vacuum, yeast growth inhibition was removed and a much higher cell density was obtained (1–1.5 times higher as compared to the process operating at atmospheric pressure). Also, the exponential growth under vacuum was extended for almost 2 h and sugar utilization was improved [32].

The continuous process when conducted under vacuum enabled rapid and complete conversion of concentrated sugar solutions. Ethanol productivity of $40\ g\ l^{-1}\ h^{-1}$ was achieved when total pressure was 50 mm Hg and glucose concentration in the feed was 33% (w/v). When continuous vacuum fermentation was performed with cell recycle, ethanol productivity was doubled. In this way ethanol productivities were increased twelvefold over conventional continuous process [10]. This improvement in productivities enables a twelvefold reduction of the required volume. As a result, investment and operating costs for industrial ethanol manufacturing are lowered.

Process operating under vacuum could be improved by the use of thermotolerant yeasts or bacteria with substrate conversion capabilities at 45 °C or above. This would permit higher pressure and decreased probability of contamination.

Ethanol productivity of a continuous process under atmospheric pressure with cell recycle was increased 4 times in comparison to the same process but without cell recycle. The productivity under atmospheric conditions with cell recycle was limited by the low feed glucose concentration, which had to be maintained to avoid ethanol inhibition. Oxygen has an important influence on viability of cells and their activity and must be present in small concentrations. In classical processes medium is supplied with oxygen by introducing a stream of air and in a vacuum system by introducing pure oxygen. The optimum oxygen feed rate for ethanol production is about 0.1 vvm. Higher oxygen concentrations decrease ethanol productivity [10,13,20]. Ethanol productivity also depends on cell concentration. This is the reason why recycling of cells is practiced in certain processes. Also, there are systems which retain cells in a fermentor with special devices such as rotating microporous membrans. This type has been claimed to replace both a classical fermentor and a centrifuge or a settler for cell separation. According to the authors, both investment cost and power consumption for the operation are about 1.6 times lower than for classical systems with cell recycle [26]. Higher pressures required for this system and preparation costs may be prohibitive for industrial applications of similar systems.

All cellulosic materials can serve as a sugar source for ethanol production. For this technology, very important steps are pretreatment and hydrolysis of these materials. The processes precede ethanol production [42,48]. This technology can be simplified by consolidating the two separate processes — hydrolysis of cellulosic materials and conversion of sugars — into one simultaneous saccharification-fermentation process which may result in a more effective and accelerated production rate [8,28,45].

Recently bacteria have been studied more thoroughly for ethanol production. It was shown that *Zymomonas mobilis* can convert glucose to ethanol efficiently and rapidly with higher specific rates of glucose uptake and ethanol production than yeasts. [53,54,55].

A new approach to ethanol production is the utilization of immobilized microbial cells [9,21,40]. Yeast cells in cross-linked gelatin can be attached to a support and then be used in a packed-bed reactor for ethanol production. In this way ethanol can be continuously produced with a concentration of about 14–15% (v/v) with residence times of 2–8 h. This reactor operated successfully for a period of several weeks [21].

Immobilized *Saccharomyces cerevisiae* (ATCC 24858) cells packed in a tubular reactor showed 9 times higher ethanol productivity than in a chemostat. Performance, stability and economy of the immobilized cell reactor was found to be superior to the conventional stirred tank for conversion of glucose to ethanol [40]. At Purdue University, Indiana, a mold is used for ethanol production from corn. By using corn starch and fiber, 20–25% increased yield of ethanol was obtained [33].

The separation of ethanol from the water-ethanol mixture is a very energy demanding process. According to Chemapec Inc. a thermocompression process in the distillation stage saves thermal energy [4]. Also, a positive energy balance is achieved when recuperated heat from aerobic and anaerobic waste water treatment (from the alcohol plant) is used in the production of alcohol from corn. When the energy from byproducts is utilized a net energy gain of 2.5×10^4 KJ l^{-1} of ethanol can be achieved [4, 33]. One way of saving energy in distillation can be through utilization of any kind of commercially available gasoline as the azeotrope [33] or mixing di-ethyl ether instead of benzene [27].

Inesco Associates Inc. (East Brunswick, N.J.) considered with Union Carbide Corp. the possibility of using molecular-sieve installations for dehydration of water-ethanol mixtures. This kind of dehydration is very effective, but the equipment is presently too expensive [33]. Use of solar energy for distillation has also been attempted [27].

4 The Economic Issues

4.1 Ethanol from Grain

The production of bioethanol from grains has to be evaluated by some key criterias which can be grouped into the following:
a) technology
b) economic and energy feasibility, and
c) general policy for energy and food.

Table 5. Cost per m^3 for producing ethanol from corn[3]

Item	Corn price per ion					
	$ 39.4	$ 59	$ 78.7	$ 98.4	$ 118.1	$ 137.8
Grain cost	100.4	150.6	203.4	253.6	303.8	356.7
Conversion cost[a]	81.9	81.9	81.9	81.9	81.9	81.9
By-product value[b]	−42.3	−63.4	−87.2	−108.3	−129.4	−150.6
Capital charge[c]						
Low	55.5	55.5	55.5	55.5	55.5	55.5
High	(84.5)	(84.5)	(84.5)	(84.5)	(84.5)	(84.5)
Cost per m^3						
Low	195.5	224.6	253.6	282.7	311.7	343.5
High (a)	(224.5)	(253.6)	(282.7)	(311.7)	(340.8)	(372.5)
High (b)	(258.9)	(287.9)	(317.0)	(346.1)	(375.2)	(406.9)

Note: Low = lowest range of both conversion and capital cost
High (a) = low range conversion cost, high capital cost
High (b) = high range of both conversion and capital cost

[a] Different studies show conversion costs range from $ 81.9 to 116.2. Expert opinion indicates $ 81.9 is theoretically possible and could be achieved in large commercial facilities

[b] Values of by-product distillers grain feed calculated at 135% of corn price. This assumption would not hold with a national gasohol program

[c] Low capital cost assumes debt financing. High includes profit margin sufficient to attract equity financing

According to today's experience, technology for the production of ethanol from grains is well established. Taking into account the general policy for energy and food, it is important to realize that in spite of shortages, constant price increases and instability of petroleum supply use of grains for food should be the first priority followed by alcohol production.

When considering the economics of alcohol production from grain, the following factors are essential:

a) cost of grain as a raw material
b) process and capital cost including labour
c) by-products credit (eg. distillers grain for feed).

The U.S. Department of Agriculture-Economics, Statistics and Co-operatives Service [3] uses corn for consideration of the economic feasibility of fuel alcohol production.

Corn is rich in starch, is the most abundant grain and is cheaper than wheat. Also, corn causes less process problems than other grains.

The cost per m^3 for producing ethanol from corn is shown in Table 5. Different grain costs, conversion cost, by-product values, low and high capital charge are considered for the calculation of ethanol costs. It is apparent from this table that the cost per m^3 of ethanol produced from corn would be (low) from $ 195.5–343.5 if the corn price is from $ 39.4–137.8 per ton or (high) from $ 244.5–372.5 (with low

Table 6. Effect of crop prices on ethanol cost [22]

	Crop price		Ethanol cost	
	Dollars/t	Dollars/bu	Dollars/L	Dollars/Imp. gal
Corn[a]	78	2.00	0.18	0.83
	118	3.00	0.27	1.24
	157	4.00	0.37	1.66
Spring wheat[b]	147	4.00	0.39	1.76
	183	5.00	0.48	2.20
	220	6.00	0.58	2.64
Winter wheat[c]	147	4.00	0.35	1.61
	183	5.00	0.44	2.01
	220	6.00	0.53	2.41
Barley[d]	092	2.00	0.24	1.07
	138	3.00	0.35	1.60
	183	4.00	0.47	2.14
Potatoes[e]	15		0.13	0.61
	100		0.90	4.09
	150		1.35	6.13
	200		1.80	8.17

Prices quoted January 11, 1980 Toronto Globe and Mail

[a] Corn: $ 114 t^{-1} (Ontario)
[b] Spring wheat: $ 201–217 t^{-1}
[c] Winter wheat: $ 173 t^{-1}
[d] Barley: $ 122 t^{-1} (Ont. feed); $ 111–124 t^{-1} (Winnipeg feed)
[e] Potatoes: $ 138–165 t^{-1} (table grade, Ontario); $ 13–20 t^{-1} (small or odd shaped culls); zero or negative value for spoiled culls plus cost of transporting)

range conversion cost and high capital cost) or from \$ 258.9 to 406.9 with high range conversion cost and high capital cost. Considering the price of the ethanol from corn at present price of \$ 114 t^{-1} in Ontario in comparison to gasoline, it is obvious that under these conditions ethanol is not competitive with gasoline.

In the above study the only by-product considered is distillers grain, i.e., the nonstarch portion of the grain that includes largely cellulosic and protein materials. As a feed it is high in protein (about 22% total digestible protein), vitamins and minerals with a high feeding value for cattle. By-product credits for fusel oil, esteraldehyde and carbon dioxide were not included. (For fuel purposes fusel oil and esteraldehyde need not be removed).

The largest percentage of the production cost of ethanol from starch crops is the substrate cost. Table 6 shows the influence of crop prices on ethanol cost for corn, spring wheat, barley, and potatoes [22]. If low grade grains are used the influence of their prices on ethanol price would be less but not much lower. "Low grade grains" refer to mechanically damaged grain. In the case of biological damage, cost for production could be even higher. This means that feedstock prices along with other obligatory production-related costs are largely affecting alcohol production economics. Production of ethanol from grains requiring governmental subsidies to surmount the difference in price between ethanol and gasoline. However, there are certain processes that claim ethanol could be produced from

Table 7. Economics of production of anhydrous ethanol from sugar cane juice [12]

Basis: 150 m^3 d^{-1} (27,000 m^3 a^{-1}) distillery of anhydrous ethanol operating 180 d a^{-1} (exchange rate: Cr \$ 15.00/U.S. \$ 1.00)

Item	Value	
Investment		
Fixed investment (10^6 U.S. \$)	13.1	
Working capital (10^6 U.S. \$)[a]	2.0	
Composition of selling price	(U.S. \$ m^{-3})	(%)
Feedstock:		
sugar cane at U.S. \$ 10.9 t^{-1}	164	49.3
Chemicals and utilities	5	1.5
By-products[b]	(16)	(4.8)
Labour	11	3.3
Maintenance materials, operating supplies insurance and administrative expenses	24	7.2
Value added taxes[c]	28	8.4
Income tax	21	6.3
Depreciation[d]	47	14.1
Net operating profit[d]	49	14.7
Calculated selling price as fuel, ex-distillery	333	100.0

[a] Includes ethanol inventory corresponding to 50 d operation at manufacturing cost
[b] Difference between the cost of direct application of stillage as fertilizer and the credit of sales of hydrated ethanol and fusel oil
[c] Major component of this item (93%) corresponds to the value of taxes levied on the feedstock
[d] Return on investment of 12% a^{-1} DCF based on the annual sum of depreciation net operating profit, and 15 years operational life for the distillery

corn at a price of 0.132–0.185 l^{-1} after taking credit for all by-products [33]. In this case ethanol would be very competitive with gasoline.

4.2 Ethanol from Sugar Cane Juice and from Cassava

De Carvalho et al. [12] have published data related to production costs of two plants for ethanol from sugar cane juice and cassava (27,000 m^3 and 49,500 m^3 anhydrous ethanol per year, respectively). Economics of the production are shown in Tables 7 and 8. According to this study ethanol prices from cassava and sugar cane juice do not differ greatly. Cost of ethanol per unit energy content was at that time significantly higher than that of gasoline. The reason is the relatively low heating value of ethanol .

It is important to mention that the cost of agricultural raw materials (cassava and sugar cane) in production has a considerable influence on the price of ethanol produced, they represent 50–60% of the ethanol price. Subsequently, the effect of lowering the cost of production of these raw materials would be appreciable.

Table 8. Economics of production of anhydrous ethanol from cassava [12]

Basis: 150 $m^3\ d^{-1}$ (49,500 $m^3\ a^{-1}$) distillery of anhydrous ethanol operating 330 $d\ a^{-1}$ (exchange rate: Cr $ 15.00/U.S. $ 1.00)

Item	External supply of electric power		Total on-site power generation
Investment			
Fixed investment (10^6 U.S. $)	15.76		19.46
Working capital (10^6 U.S. $)	1.06		1.08
Composition of selling price	(U.S. $ m^{-3})	(%)	(U.S. $ m^{-3})
Feedstock			
Cassava roots at U.S. $ 29.2 t^{-1}	200	59.1	200
Enzymes and chemicals	29	8.6	29
Utilities			
Water	—		—
Electric power at U.S. $ 28.5 $MW^{-1}\ h^{-1}$	13	7.4	—
Wood at U.S. $ 7.0 t^{-1}	12		31
By-products' credit[a]	(18)	(5.3)	(20)
Labour	9	2.7	9
Maintenance materials, oper. supplies, insurance & administrative expenses	17	5.0	20
Value added taxes[b]	3	1.0	4
Income tax	12	3.5	16
Depreciation[c]	32	9.4	38
Net operating profit[c]	29	8.6	36
Calculated selling price as fuel, ex-distillery	338	100.0	363

[a] Difference between the cost of direct application of stillage as fertilizer, and the credit of sales of hydrated ethanol and fusel oil

[b] Value of social tax, only. Feedstock is considered exempt of taxes and does not contribute to this item

[c] Return on investment of 12% a^{-1} DCF. based on the annual sum of depreciation net operating profit and 15 years operational life for the distillery

Table 9. Required fixed capital investment (in 10^3 dollars) and ethanol production costs* (in $ m^{-3}) for different modes of operation [11]

	Plant capacity was 295 m^3 d^{-1} 95% ethanol			
	Batch	Continuous	Continuous cell recycle	Vacuum-cell recycle
Fermentation	14,900 (44.40)	4,800 (17.44)	2,484 (13.21)	3,366 (12.68)
Ethanol recovery	928 (21.40)	928 (21.40)	928 (21.40)	824 (17.70)
Yeast recovery	962 (5.28)	962 (5.28)	1,362 (6.34)	794 (2.64)
Storage	811 (1.58)	811 (1.58)	811 (1.58)	233 (0.53)
Total	17,601 (72.66)	7,509 (45.70)	5,585 (42.53)	5,217 (33.55)
Sugar			(194.71)	(188.64)
Medium supplements			(15.32)	(7.13)
Total production costs			(252.56)	(229.32)
Yeast credit			(35.14)	(16.38)
Net production costs			(217.42)	(212.94)

* Ethanol production costs are shown in brackets

The capital investment for sugar cane and cassava distilleries of the same capacity are of the same order of magnitude. Equipment for conversion of starch to sugar and electric power generation influenced higher initial fixed investment in cassava distilleries. In the process of alcohol production from sugar cane according to Brazil legal requirements are that the production plant must have ethanol storage capacity adequate for a continuous ethanol supply throughout the whole year including the sugar cane off-season. This requires an increase in investment in plants for sugar cane processing.

4.3 Ethanol from Molasses

Cysewski and Wilke [11] have conducted an economic analysis of ethanol production from molasses on the basis of preliminary process design studies both for continuous and batch technologies. Designs are based on a production capacity of 295 $m^3\ d^{-1}$ of 95% ethanol. Fixed capital investments and ethanol production costs for different modes of operation are shown in Table 9.

The highest fixed capital investment is required for the batch process and the lowest for vacuum-cell recycle operation. Fixed capital investment for vacuum-cell recycle operation is only 30% of that for the batch operation. The ethanol production costs are also the highest for the batch process. The operating costs and the return on investment are presented in Tables 10 and 11, respectively.

The influence of investment related costs on the production costs are very high, especially in a batch process. In this case production costs are lower for classical continuous processes or continuous processes with cell recycle than for the batch process, and the lowest cost is for the vacuum-cell recycle operated process.

For this analysis the cost of molasses was taken at 11 cents kg^{-1} sugar, and it is supposed that 97% of the sugar is processable. From the data presented, it is obvious that the sugar price has the greatest influence on the total production cost

Table 10. Operating costs for different processes for ethanol production* [11]

	Production cost ($ m^{-3})			
	Batch	Continuous	Continuous cell recycle	Vacuum cell recycle
Investment related costs	27.21	12.94	10.57	9.25
Operating labour	8.45	2.38	1.32	1.06
Supervision and clerical	0.53	0.26	0.26	0.26
Utilities				
Water	1.58	1.58	1.58	1.06
Power	3.17	1.58	2.38	1.58
Steam	26.68	25.10	25.10	17.96
Oxygen	—	—	—	1.32
Laboratory changes	0.26	0.26	0.26	0.26
Plant overhead	4.75	1.58	1.06	0.79
Total	72.63	45.68	42.53	33.54

* Plant capacity was 295 $m^3\ d^{-1}$ 95% ethanol from 50% "cane" molasses sugar solution

Table 11. Comparison of return on investiment before taxes between the various processes* [11)]

Processes	Percent return on investment[a)]	
	Yeast credit taken at 22 cent kg^{-1}	No yeast credit
Batch	18.5	3.3
Continuous	55.6	27.9
Continuous with cell recycle	69.7	36.5
Vacuum with cell recycle	81.5	64.3

* Selling price of 95% ethanol taken as \$ 290.6 m^{-3}
[a] Percent return on investment — (yearly profit/total capital investment) × 100

of ethanol and then even a slight increase in sugar price causes a significant escalation in the ethanol price. The total production cost of ethanol can be reduced for the yeast credit. In this analysis the yeast credit was taken at 22 cent kg^{-1}. After subtraction of the yeast credit, net production cost for ethanol seems to be almost the same for continuous cell recycle and the vacuum cell recycle operations. However, if all aforementioned processes are compared with respect to return on investment (Table 15), it is clear that the greatest return is obtained with the vacuum with cell recycle process. A particular advantage of that process in relationship to the others is observed when the yeast credit is not considered. The lowest return on investment is obtained in the batch process. As the batch process is the conventional technology for ethanol production, its low return on investment is perhaps a reason why more bioethanol has not been produced. From that point of view, new technologies could influence an increase in ethanol production.

4.4 Ethanol from Urban Wastes

This economic analysis is based on ethanol production from urban waste through its enzymatic hydrolysis following conversion resulting sugars [42)]. A plant with a capacity

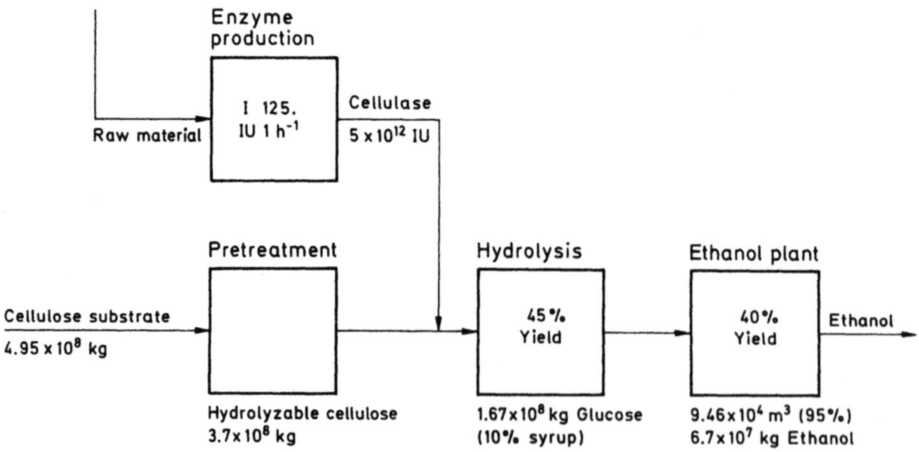

Fig. 2. Overal material balance for production of ethanol from urban waste (Amounts per year) [42)]

Table 12. Operating cost for production of 95% ethanol from urban waste* [42)]

	Enzyme production	Pre-treatment	Hydrolysis	Ethanol production	Total
Total raw material	50.99		0.53	0.10	51.62
Total material[a)]	57.86	1.64	3.46	5.07	68.03
Total utilities[b)]	15.67	36.99	6.00	22.30	80.95
Variable cost[c)]	73.53	38.63	9.46	27.37	148.98
Total direct labour[d)]	12.81	7.58	7.69	12.10	40.18
Total direct cost[e)]	86.34	46.21	17.15	39.47	189.16
Plant overhead	10.25	6.29	6.15	8.93	31.62
Tax and insurance	3.99	1.11	1.85	4.07	11.02
Depreciation	19.97	5.49	9.27	20.32	55.06
Total	120.55	59.10	34.42	72.79	286.86
Cellulose substrate					34.87
Factory cost[f)]					321.73

[a] Raw material, maintenance material, and operating supplies
[b] Utilities, cooling water, steam, process water, and electricity
[c] Total materials and total utilities
[d] Operating labour, maintenance labour, and control laboratory
[e] Variable cost and total direct labour
[f] Total direct cost plus overhead, tax, insurance, and depreciation
* Cost expressed in $ m^{-3}

of 9.46×10^4 m^3 a^{-1} of 95% fuel grade ethanol has been considered for the cost analysis on the basis of laboratory experimental data. A material balance and process flow diagram are shown in Fig. 2. The operating and capital costs for this plant are presented in Tables 12 and 13, respectively. As shown, the factory cost of ethanol from urban waste is 32 cent per liter including 3.5 cent per liter for the cost of substrate. In the ethanol price enzymes constitute 37.5%, ethanol production

The above costs were calculated on the following basis:

Depreciation	— 10% of total fixed capital
On-stream factor	— 0.90; 330 d a^{-1} (7,920 ha^{-1})
General administration, Sales and research profit	— not included in factor cost
Maintenance (material and labour)	— 3% a^{-1} of battery limit investment plus 1% a^{-1} of refrigeration investment
Operating supplies	— 10% of operating labour cost
Operating labour cost	— $ 11.65/man-h
Plant overhead	— 80% of total labour cost
Taxes and insurance	— 2% a^{-1} of the total fixed capital
Laboratory	— 20% of operating labour cost
Services	— 10% of the sum of the battery limits investment and off-site investment including 1% allowance for waste handings
Battery limit investment	— includes all major equipment
Off-site investment	— includes off-site utilities and tankage investment plus 10% contingency

Table 13. Capital cost for production of 95% ethanol from urban waste* [42]

	Enzyme production	Pre-treatment	Hydrolysis	Ethanol production	Total
Total plant equipment	7,048	1,750	2,640	4,552	15,990
Battery limit investment	15,506	3,500	7,920	13,656	40,582
(factors)	(×2.2)	(×2)	(×3)	(×3)	
Off-site investment	1,681	1,234	64	3,815	6,794
Subtotal, battery limit and off-site investment	17,187	4,734	7,984	17,471	47,376
General services facilities (10% BL & OI)	1,719	473	798	1,747	4,737
Total fixed investment	18,906	5,207	8,782	19,218	52,113
Start-up (8.5% TFI)					4,430
Working capital (16.5% TFI)					8,599
Total capital investment					65,142

* Units in U.S. $ 1,000. 1979)

(fermentation and distillation) 22.6%, pretreatment 18.4%, substrate 10.8% and hydrolysis 10.7%.

The factory cost of ethanol can be reduced by 23% if credits are taken for process steam and cellular biomass. The residual cellulose from the hydrolysis plant can produce process steam with an estimated credit of $ 26.42 m^{-3} of ethanol as based on the combustion energy value of 1.39×10^4 kJ kg^{-1} or $ 2.61/10^6 KJ. Cellular biomass is a by-product from the ethanol plant with an estimated credit of $ 47.56 m^{-3}. This biomass may be used as animal feed and/or fertilizer at a market value of $ 150 t^{-1}. Taking the above two credits into consideration the factory cost of ethanol would be reduced to $ 248.35 m^{-3}.

4.5 Ethanol from Wood

Wayman et al. [50] have given rough estimates of production costs of ethanol from wood. They have taken into consideration processes for acid and enzymatic hydrolysis of aspen wood. The product is 95% ethanol at a plant capacity of 4.55×10^4 m^3 a^{-1}, representing according to these authors, a minimum eçonomic size. Tables 14 and 15 show capital and annual operating costs, respectively.

According to this analysis the cost of one m^3 of ethanol obtained through acid hydrolysis of aspen wood would be $ 312.1 and that obtained through enzymatic hydrolysis would be $ 295.2. Cost of distribution and profit are included in these prices but not tax. If the price of ethanol obtained through enzymatic hydrolysis is compared with that obtained through acid hydrolysis the former is lower by about 6%. Capital cost is also 10% lower for a plant assuming enzymatic hydrolysis. (The authors state that these prices are competitive with industrial alcohol, but not with today's gasoline prices). A material balance of ethanol production through enzymatic

Table 14. Capital costs for production of ethanol from wood* (in \$) [50]

	Acid	Enzyme
Wood requirement, chips, tons d^{-1}	444.5	376.5
Capital costs, (million \$)		
Chip storage and distribution	1.0	0.85
Autohydrolysis stage	4.0	3.5
Caustic extraction stage	1.0	1.0
Hydrolysis stage	4.0	2.0
Conversion	2.8	3.5
Distillation	1.8	1.8
Steam system	1.6	0.8
Water system	0.15	0.4
Material storage and handling	1.0	1.0
Pollution control	0.15	0.15
Buildings and land	3.5	3.5
Total physical plant	21.0	18.5
Engineering construction, overhead, contigency	4.0	4.0
	25.0	22.5
Working capital	3.0	3.0
	\$ 28.0	\$ 25.5

* Capacity 4.55×10^4 m^3 a^{-1} of 95% ethanol

Table 15. Operating costs for production of ethanol from wood* (in million \$) [50]

	Acid	Enzyme
Annual operating costs \$		
Wood		
Caustic soda	4.3	3.63
Sulphuric acid	0.2	0.2
Chemicals	0.6	1.0
Packaging and distribution	1.0	1.0
Total variable costs	6.1	5.83
Utilities	0.1	0.1
Labour	2.0	2.0
Maintenance materials	0.2	0.2
Other	0.2	0.2
Total fixed costs	2.5	2.5
Total annual operating costs	8.6	8.33
20% of investment	5.6	5.1
Total annual costs	\$ 14.2	\$ 13.43
Price per 1 m^3 of ethanol, including distribution and profit, but no tax	\$ 312.1	\$ 295.2

* Capacity 4.55×10^4 m^3 a^{-1} of 95% ethanol

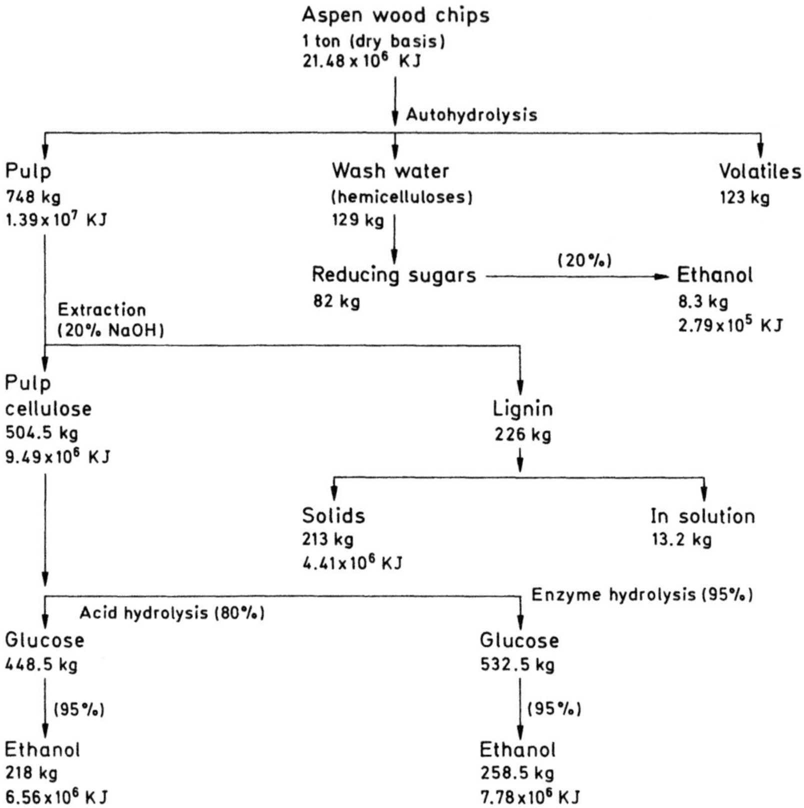

Fig. 3. Material and energy balance for ethanol production from wood, by the acid and enzymatic hydrolysis [50]

and acid hydrolysis is also presented (Fig. 3). As a starting material for the production, 1 ton of dry aspen wood chips were used.

The major factor which enters into the variable operating cost is wood. In this analysis it is taken as \$ 27.55 t^{-1}. If waste aspen chips were used (\$ 19.84 t^{-1}), the ethanol price would be cut by \$ 22–26.4 per m^3. With government subsidies, capital cost could be reduced by 50%; this would result in a lowering of the ethanol price by \$ 55 to \$ 61.6 per m^3. The authors estimate that ethanol could cost 22 cent l^{-1} if the project were properly located and funded. Increase in the plant capacity could also reduce the price of ethanol.

4.6 Ethanol from Wheat Straw

SRI International under contract from the U.S. Department of Energy [43] have conducted an economic feasibility study of a process for converting wheat straw to fuel grade ethanol through enzymatic hydrolysis. This analysis is based predominantly on the studies of Wilke's group, University of California, Berkley. Some data are also taken from other sources such as from the U.S. Army Natick Development Center at Natick, Massachusetts. Many of the design assumptions have not been fully tested because the research used as a basis of the process design were only made through bench-scale development.

The basis for this economic analysis is a plant with a capacity of 9.46×10^4 m^3 of 95% ethanol per year. The technological process is divided into the following 6 operating sections: substrate pretreatment, enzyme recovery, hydrolysis, sugar solution concentration, enzyme production and ethanol production. It is assumed that all sections operate continuously except for the conversion of the sugar to ethanol. From the preliminary design prepared by SRI, an estimate was made of the required capital investment for the base case (9.46×10^4 m^3 a^{-1} of 95% ethanol) constructed in the U.S.A. in December 1977 as follows:

	$ Million
Battery limits investment by plant section	
Substrate pretreatment	12.0
Enzyme recovery	3.5
Hydrolysis	19.2
Sugar solution conc.	4.4
Enzyme production	21.3
Sugar conversion and alcohol distillation	10.1
Subtotal	70.5
Off-sites, including utilities, tankage, waste treatment and general service facilities	26.3
Total fixed capital	96.8
Startup cost	8.2
Working capital	15.9
Total capital investment, excluding land	120.9

The enzyme production and hydrolysis plant sections represent the most costly plant sections, accounting for 31% and 27% of the fixed capital investment respectively. The plant gate manufacturing cost for the production of 9.46×10^4 m^3 a^{-1} of 95% vol. ethanol is summarized as follows:

	$ m^{-3}
Total materials cost	538.9
Utilities cost	126.8
Capital cost	121.5
Labour cost, including overhead	95.1
Plant gate manufacturing cost[a]	882.3

[a] Plant gate cost includes raw materials, utilities, maintenance material and labour, operating labour and supplies, control laboratory, plant overhead, taxes, insurance, and depreciation but does not include any general and administrative, sales, or research costs, profit, or any by-product credits or disposal charges.

For a 15% discounted cash flow rate of return on investment after taxes, the selling price for the 95 vol% ethanol would have to be $ 1.29 l^{-1}. This is based on a 15% a^{-1} of fixed capital for general and administrative, sales and research costs, a

15 year expected life of the plant, a 48% tax rate and capacity production throughput, requiring no sales buildup over the years.

4.7 Ethanol from Cornstalks

Production of ethanol from cornstalks was also taken into consideration [39,40]. After acid hydrolysis xylose and glucose were obtained. Glucose is converted to ethanol by *S. cerevisiae* in an immobilized cell reactor. *Fusarium oxysporum* converts xylose to ethanol in another immobilized cell reactor. Cell overgrowth is removed every two weeks by sparging the reactors with compressed carbon dioxide. The ethanol in the reactor effluent is separated by distillation in a 35 stage column. Reactors were designed for 95% conversion of the sugar. When the conversion decreases to 90% the reactors could be regenerated. A plant for conversion of cornstalks to ethanol for economic viability would have a minimum capacity of 1.7×10^4 m^3 of ethanol per year. The capital investment costs with a 30% contingency are shown in Table 16. Almost 50% of the capital investment is for the acid hydrolysis process and for acid recovery. The operating cost for the process is shown in Table 17. In this economic analysis the price of corn residue is estimated as being \$ 25 t^{-1}. Under this condition, raw material constitutes 36% of the operating cost. The breakeven ethanol price is \$ 298.55 m^{-3}. (Current 1980 selling price is \$ 383.1 m^{-3}.) The authors estimate that with a four-fold increase in plant size the breakeven price could be reduced to \$ 264.2 m^{-3}. If corn residues cost \$ 15 t^{-1}, the breakeven price of ethanol would be \$ 237.8 m^{-3}. These projected economics of ethanol production from corn residues look quite attractive. The authors conclude that the production of ethanol for use as a petrochemical would be feasible at today's prices. Indeed ethanol produced by this process would be used as a fuel at gasoline prices up to approx. \$ 330 m^{-3}.

Table 16. Capital investment for ethanol production from cornstalks [40]

Item	Dollars
Hydrolysis reactors (2)	145,000
Acid recovery	2,472,000
Pumps and piping	253,300
Heaters	203,500
Rotary vacuum filters (2)	394,000
Rotary dryer	277,700
Impregnator	334,500
Reactors (2)	512,100
Compressor	5,200
Distillation column	670,200
Centrifuges (2)	389,900
Storage tanks	20,300
Heat exchangers	113,200
Contingency (30%)	1,737,270
Total	7,528,170

Table 17. Operating costs for ethanol production from cornstalks [40]

Item	Dollars per year
Cornstalks	1,825,000
Acid recovery	300,700
Make up acid	421,700
Utilities	381,200
Neutralizer	265,000
Yeast extract	256,000
Labour	409,400
Maintenance	409,400
Depreciation, taxes, insurance	818,800
Total	5,087,200
Production (m^3 a^{-1})	1.7×10^4
Breakeven price (\$ m^{-3})	298.55

5 The Energy Issues

5.1 Ethanol from Corn

For production of 35.24 l (1 bushel) of corn, 1.13×10^5 KJ of energy is required. If the bushel of corn is directly combusted 3.95×10^5 are recovered. If the bushel of corn is fermented by yeasts it will be converted to:

ethanol	9.84 l — if combusted yield	2.30×10^5 KJ
distillers grains	7.7 kg — if combusted yield	1.37×10^5 KJ
carbon dioxide	7.4 kg — not combustible	— —
		3.67×10^5 KJ

and slightly less energy will be obtained than the raw corn contains. Distillers grains are not combusted, they are used for animal feed. Conversion of corn to ethanol requires 3.57×10^5 KJ to 4.15×10^5 KJ per 35.24 l. By adding the amount of energy used for corn production, total energy expenditure is 4.70×10^5 to 5.28×10^5 KJ per 35.24 l. It is obvious that only 0.69 to 0.78 KJ of energy (ethanol and distillers grains) is produced from each KJ of energy used in the growth of corn and its subsequent conversion to ethanol. If we consider that only starch is converted to ethanol and that distillers grains which are recovered have a beef feeding value equivalent to 14.45 l of corn, it means that only 59% of the energy is used to produce a bushel of corn (6.68×10^4 KJ). In this situation, for each KJ used for corn and ethanol production ethanol representing only 0.48 to 0.54 KJ is obtained. From this calculation it is clear that the overall energy balance for ethanol production from corn is negative [3].

Table 18. Energy consumption and production in corn farming [36]

Item	KJ per 35.24 l (1 bushel) corn	KJ per 3.785 l (1 gallon) EtOH
Energy consumption in corn farming		
Seed corn	3,072	1,185
Fertilizer	54,492	21,006
Herbicide and insecticide	1,135	437
Gasoline	39,241	15,128
Electricity	15,998	6,167
Irrigation	1,734	668
Labour	253	98
Drying	6,193	2,387
Transportation of corn	3,612	1,193
Total	125,730	48,269
Energy content of products from corn farming		
Corn[a)]	360,204	128,314
Stalks, cobs, husks[b)]	453,354	174,768
Total	813,558	303,082

[a] Digestible energy; [b] Lower heating value

Table 19. Energy consumption and production in the manufacture of ethanol from corn [36]

Plant section	KJ per 35.24 l (1 bushel) corn	KJ per 3.785 l (1 gallon) EtOH
Energy consumption in an ethanol plant		
Liquid concentration	119,228	45,965
By-product grain drying	53,607	20,669
Subtotal, by-products	172,835	66,634
Grinding, cooking, propagation	66,065	25,465
Fermentation	1,539	590
Beer still and centrifuge	110,491	42,603
Distillation	78,101	30,102
Dehydration	38,882	14,988
Subtotal, ethanol	295,078	113,748

Item	KJ per 35.24 l (1 bushel) corn	KJ per 3.785 l (1 gallon) EtOH
Energy content of products from an ethanol plant		
Ethanol[a)]	206,698	79,682
Aldehydes, fusel oil[a)]	2,953	1,138
By-product grains[b)]	123,170	47,483
	332,821	128,303

[a] Lower heating value; [b] Digestible energy

Other economic analyses for energy input and output for ethanol production from corn are also available. One such analysis is presented in Table 18 showing the energy consumption in corn farming and energy content of products from farming [36]. The energy consumed in the plant and energy content of obtained products is shown in Table 19.

In summary, the consumed energy for corn and ethanol production and that obtained through products is shown in Table 20. It is clear that the overall energy balance is positive and that the net energy production per 3.785 l (1 gallon) of ethanol is 91,727 KJ. Not subtracted from that value is the energy spent for production of distillers dried grains and solubles because this process is not directly related to the production of ethanol.

The positive energy balance was mainly achieved by taking into account the energy that could be obtained and used in the process through burning of the whole quantity of stalks, cobs and husks. Scheller and Mohr [36] proposed for practical reasons to use about 75% of cobs, stalks and husks and to leave the rest in fields for soil conditioning. In that case the net energy production is still about 48,000 KJ per gallon of ethanol if the energy deficit for by-product production is not included or more than 28,500 KJ per gallon of ethanol if that deficit is included.

This analysis is one of the most optimistic. Removing the stalks, cobs and husks creates a problem because it means mining of soil, and in a short period of time it would cause environmental problems.

Table 20. Overall energy balance for grain alcohol production from corn [36)]

Energy production	KJ per 35.42 l (1 bushel) corn	KJ per 3.785 l (1 gallon) EtOH
Ethanol	206,698	79,682
Aldehydes, fusel oil	2,953	1,138
Stalks, cobs, husks	453,354	174,768
Total	663,005	255,588
Energy consumption		
Farming operation	125,731	48,469
Transportation of stalks, etc.	4,265	1,644
Alcohol plant	295,078	113.748
Total	425,074	163,861
Net energy production	237,932	91,727
Net energy loss		
By-product grain production	49,664	19,151

From the other point of view, use of these materials as a source of energy in the process of ethanol manufacturing has not yet been solved. But in spite of all, if that material is burned the term "renewable materials" would be in question.

If we do not accept the above analysis in its entirety a negative energy balance that appears if stalks, cobs and husks are not used for burning in the process may possibly become less negative or even slightly positive with new, more economical technology for growth of grains and for ethanol production from them.

Also, it would be favourable to consider the energy contained in the main by-product — distillers grain. This material is not burned and consequently no cumbustion heat is obtained. This by-product is used for animal feeding, and it allows less energy to be spent for production of another kind of animal feed that would be replaced by distiller's grains. That part of saved energy should be also attributed to the produced energy in ethanol production. A considerable amount of energy is also contained in methane obtained through anaerobic treatment of waste waters of that industry. In this way a more favourable energy balance in production of ethanol could be achieved [4)].

Taking into account that the energy crisis is caused by the shortage of crude oil it is interesting to compare how much of petroleum energy is used for growing corn and for ethanol production. These data are shown in Table 21 [38)].

Three different processes are compared. For each process the petroleum type energy input is higher than output through ethanol and for each liter of ethanol produced more than 1 liter of petroleum energy equivalent would be used. But if it is subtracted petroleum type energy used for by-products recovery from total petroleum type energy input in that way petroleum type energy input would be equal or slightly lower than energy obtained from ethanol produced. Bearing in mind the shortage of crude oil, it would be preferable to replace petroleum type of energy in the process of ethanol production as much as possible with other energy forms and especially with a waste energy if it is disposable.

Table 21. Use of petroleum type energy for growing corn and for alcohol production [38]

Input	Today $KJ\,l^{-1}$	In one year [a] $KJ\,l^{-1}$	Under study[b] $KJ\,l^{-1}$
To grow corn	9,688	9,688	9,688
To cook and convert	3,990	3,990	3,990
Germ recovery	624	624	624
Distilling	7,906	7,906	5,534
Gluten recovery	731	731	731
Feed recovery[c]	7,303	4,416	4,416
Total electrical energy	1,061	1,061	1,061
Total per liter	31,303	28,416	26,044
Subtract electrical energy[d]	1,061	1,061	1,061
Petroleum type energy input	30,242	27,355	24,983
Output[e]			
Alcohol (If LHV)	21,052	21,052	21,052
(If HHV)	23,391	23,391	23,391
Net balance for this plant			
(If LHV)	-9,190	-6,303	-3,931
(If HHV)	-6,851	-3,964	-1,592
Liters petroleum energy equivalent used for each liter of alcohol produced			
(At LHV)	1.436	1.299	1.187
(At HHV)	1.293	1.169	1.068

[a] By using a recompression evaporator with an electric motor
[b] By using ether for dehydrating
[c] If stillage is fed to livestock direct from the distillery, then some of the energy would not be used for feed recovery. However, this alcohol plant does use the energy and since it is not available for other users, it must be included for a proper comparison
[d] Although recompression equipment is to be powered by electricity, no allowance was made by the distiller for increased use. However, electricity was assumed to be generated using nonpetroleum type fuels (coal or nuclear) although a portion or all may be generated using oil or natural gas
[e] High values (HHV) were used throughout the application which tends to maximize the KJ value of the alcohol output. The low heating value (LHV) is also presented here to provide another comparison

5.2 Ethanol from Sugar Cane and Cassava

Ethanol produced from sugar cane or from cassava is a process greatly used in countries such as Brazil. In Tables 22 and 23 energy input and output are shown for production of 1 m^{-3} of anhydrous ethanol from sugar cane and cassava in plants with a capacity of 150 m^3 of ethanol per day [12]. A 150 m^3 of ethanol per day distillery was chosen since this size represents an average capacity of a distillery considering the Proalcool approved projects in 1977.

Taking into consideration the energy balance for ethanol production from sugar cane it is supposed that total electric energy is generated on site and sugar cane bagasse is used as a fuel for steam generation. In the process with cassava,

Table 22. Energy inputs and outputs for sugar cane farms/distillery system [12]

(Total on-site generation of electric power)
Basis: 1 m^3 anhydrous ethanol
150 m^3 d^{-1} ethanol distillery

Flows crossing system boundaries	Consumption or production		Equivalent energy (10^3 kcal)
	Unit	Quantity	
Inputs			
Distillery			
Benzene	kg	0.90	8.73
Ammonium sulphate	kg	22.55	4.51
Superphosphate	kg	22.70	4.09
Labour + others	—	—	nil
Sugar cane farms			
Diesel	kg	37.34	373.40
Ammonium sulphate	kg	121.00	24.20
Potassium chloride	kg	24.50	18.50
Superphosphate	kg	10.50	1.89
Herbicides	kg	0.69	0.12
Insecticides	kg	0.11	0.02
Labour	man-day	7.55	0.02
Farms-distillery transportation			
Diesel (stillage)	kg	12.77	127.70
Diesel (sugar cane)	kg	9.73	97.30
Total inputs	—	—	660.48
Outputs			
Anhydrous ethanol	kg	790.00	5,271.67
Hydrated ethanol	kg	48.00	306.38
Fusel oil	kg	2.00	12.77
Total outputs	—	—	5,590.82

external electric power is used and wood is the fuel source for process steam generation

In Table 24 the energy balance and net energy ratio (output/input) are shown comparatively for production of ethanol from sugar cane, cassava with external supply and cassava with internal generation of electric power. It is obvious from this table that the energy output is almost 6–9 times higher than the input.

5.3 Ethanol from Wood

If we consider the experimental production of ethanol from aspen wood chips [50], following autohydrolysis and caustic extraction the ratio of output/input energy is 3.2 and 3.7 for a process assuming acid hydrolysis and enzymatic hydrolysis respectively (Table 25).

It can also be seen from the energy balance that all energy required to operate the process can be supplied by the lignin that is recovered. For this calculation the authors assumed a 95% yield of ethanol from wood sugars obtained after hydrolysis.

Table 23. Energy inputs and outputs of cassava farms/distillery system [12]

(Forest included in the system boundaries, external supply of electric power)
Basis: 1 m^3 anhydrous ethanol
150 m^3 d^{-1} ethanol distillery

Flows crossing system boundaries	Consumption of production		Equivalent energy (10^3 kcal)
	Unit	Quantity	
Inputs			
Distillery			
Electric power	kWh	448.00	385.23
Sodium hydroxide	kg	2.93	21.10
Benzene	kg	0.90	8.73
Molasses	kg	5.66	6.51
Ammonium sulphate	kg	22.55	4.51
Superphosphate	kg	22.70	4.09
Labour + others	—	—	nil
Cassava farms			
Diesel	kg	27.42	274.20
Ammonium sulphate	kg	63.00	12.60
Potassium chloride	kg	15.00	11.33
Insecticides	kg	9.96	1.79
Superphosphate	kg	7.50	1.35
Electric power	kWh	0.59	0.51
Labour	man-day	47.50	0.14
Farms & forest-distillery			
Transportation			
Diesel (stillage)	kg	12.26	122.60
Diesel (cassava roots)	kg	5.48	54.80
Diesel (wood)	kg	3.54	35.40
Total inputs	—	—	944.99
Outputs			
Anhydrous ethanol	kg	790.00	5,271.67
Hydrated ethanol	kg	41.60	265.53
Fusel oil	kg	4.80	30.64
Total outputs	—	—	5,567.84

It is also useful to mention that the effectiveness of acid hydrolysis to supply glucose is 80% while enzymatic hydrolysis is 95% efficient (Fig. 3).

5.4 Ethanol from Cornstalks

The energy requirement and energy in the process for ethanol (95%) production through acidic conversion of cornstalks to sugars and then through fermentation and distillation are presented in Table 26 [39]. A production of 1,708 kg (572 gal) of ethanol per h is taken as a basis for that energy balance. A positive energy balance (as presented in that table) was obtained assuming that almost all necessary energy can be supplied from fermentors (cooling) and from remaining solids. A net energy yield of this process is the ethanol produced (2,165 l per h or 1.26×10^9 KJ per d).

Table 24. Energetics of ethanol production from sugar cane and cassava [12]

Basis: 1 m^3 anhydrous ethanol

Raw material	Case	Energy (10^6 kcal)					Net energy ratio
			Input				(Output/ Input)
		Output	Agri-culture	Distillery	Trans-portation	Total	
Sugar cane	Total on-site generation of electric power (sugar cane bassage as fuel)	5.59	0.42	0.017	0.26	0.70	8.0
Cassava	External supply of electric power (wood as fuel)	5.57	0.30	0.43	0.21	0.94	5.9
Cassava	Total on-site generation of electric power (wood as fuel)	5.57	0.30	0.045	0.27	0.62	9.0

6 Summary and Conclusions

The shortage of crude oil and consequently its increasing prices cause an extensive research for its supplementation. In spite of the shortages according to estimates for the year 2000 crude oil will still constitute 37% of the world's energy supply.

Bearing in mind the special features of fuels from curde oil new energy sources have to possess similar characteristics. A possible portion supplement of crude oil fuels particularly in motive-end use can be ethanol in spite of its lower heating value. Ethanol is a natural fuel and is a readily available liquid that can be produced and utilized within existing technologies. No further processing is required for its oil use as a fuel. It burns cleaner and causes less environmental problems than petroleum. Due to biodegradibility of ethanol, handling of spills is easier. Miscibility of ethanol with water is also beneficial in this respect. This fact enables preparation of gasoline-ethanol blends resulting in gasohol for internal combustion engines. Gasohol has a higher octane value and allows both better compression ratio and energy efficiencies in comparison to gasoline. Engines operating on gasohol start easier at cold temperatures which is of advantage in countries with cold climates.

Production of ethanol by microbial processes is an attractive alternative as renewable materials such as sugars, starches, agricultural and forestry cellulosic materials and both urban and industrial wastes can be used as raw materials for its production. Some of these materials are available in large quantities. They can be used for ethanol production permanently if they are managed properly. Some of these materials are ready for conversion to ethanol and others need special pretreatment and hydrolysis. Improvements in pretreatment processes and particularly in enzyme

Table 25. Energy balance for conversion of aspen wood chips to ethanol [50]

Basis: 1 ton of aspen wood chips (dry)
Heat of combustion of aspen chips: 21.48 × 10⁶ KJ
Gross energy recovery:

	KJ × 10^6	
	Acid hydrolysis	Enzyme hydrolysis
Ethanol	6.84	8.06
Lignin	4.41	4.41
Volatiles	+	+
Total:	11.25	12.47
Energy recovery:	52.35%	58.03%
Energy required:		
Autohydrolysis	0.70	0.70
Caustic extraction	0.12	0.12
Hydrolysis	0.70	0.23
Distillation	1.97	2.32
Total:	3.49	3.37
Net energy recovery:	36.1%	42.3%

Table 26. Energy balance for conversion of cornstalks to ethanol [39]

Item	Heat available, $10^6 \times KJ\,h^{-1}$	Temperature	Heat required, $10^6 \times KJ\,h^{-1}$
Acid mixing in first hydrolysis tank	5.80	100 °C	—
Acid reaction in impregnator	1.84	80 °C	—
Acid mixing in second hydrolysis tank	1.42	110 °C	—
Feed to second hydrolysis	—	80—110 °C	5.80
Cooling feeds to fermentors	56.39	110—25 °C	—
Heating feeds from fermentors	—	25—90 °C	48.90
Heat supplied to reboiler	—	100 °C	35.20
Heat recovered from condenser	7.38	84 °C	—
Energy from remaining solids	39.00	—	—
Energy required for pumps and centrifuges	—	—	0.84
Totals	111.83		90.74

technology will have an impact on the competitive industrial production of fuel ethanol in bio-processes such as continuous process without or with cell recycle, process operating under vacuum, simultaneous saccharification-ethanol biosynthesis, utilization of immobilized cells etc. enables higher yield of ethanol and reactor productivities and a more complete use of the substrate. All these factors reflect favourably on the ethanol price.

Possible application of thermocompression, various azeotropes, molecular sieves, solar and waste heat energy are promising for the improvement of the ethanol recovery economy. These improvements are predominantly achieved in the laboratory or pilot plant scale but have not yet contributed to large scale production of low-price ethanol that could be competitive with gasoline.

High prices of "classical" raw materials for ethanol production by microbial processes are a major factor influencing ethanol production costs. Less costly cellulosic and waste materials require additional processing that is also directly related to the production costs. Substantial efforts are required to improve the economy of production of raw materials.

Overall energetics constitute a decisive factor in terms of ethanol production by microbial processes. Of particular importance is the ratio of energy required to produce the raw material and convert it to ethanol to the energy available in the final product(s). This ratio seems to be very favourable in case of sugar cane and cassava when all the bagasse is being combusted and the energy used in the process. A positive ratio is also obtained when wood chips or agricultural residues are considered as raw materials.

Considering present technology for grain production and subsequent conversion to ethanol, the energy input is greater than the output. Only with improved processes and bio-systems can this be made more favourable.

In addition to using ethanol as a fuel, it has a considerable value as a chemical feedstock. A new organic chemical industry can emerge from ethanol resulting in production of "alkochemical" as opposed to today's "petrochemicals".

7 References

1. Aiba, S., Shoda, M., Nagatani, M.: Biotech. Bioeng. *6*, 845 (1968)
2. Allsup, J. R., Eccleston, D. B.: Ethanol/gasoline blends as automotive fuel. Bartlesville Energy Technology Center, Bartlesville, Oklahoma. U.S. Govt. Reports and Announcements, V. 79, No. 22, p. 198, 1979
3. Anonymous: Gasohol from grain — the economic issues. Prepared for the Task Force on Physical Resources Committee of the Budget U.S. House of Representatives by Economics, Statistics and Cooperatives Service, U.S. Department of Agriculture, Washington, D.C. 620.8 U 58/39 ESCS11, 1978
4. Anonymous: Power alcohol: A publication by PEC (Process Engineering Company), Männedorf, Switzerland 1979
5. Anonymous: Faster growth seen for nonoil energy. Oil and Gas J. *78* (15), 56 (1980)
6. Anonymous: The administration will stick to an ambitious goal for increasing gasohol output. Chem. Eng. *87* (2), 62 (1980)
7. Block, F., Brown, G. F., Farlas, D. F.: Am. Potato J. *50* (10), 357 (1973)
8. Brooks, R. E., Su, T., Brenan, M. J., Jr., Frick, J., Lynch, M.: Bioconversion of plant biomass to ethanol. Prepared for U.S. Dept. of Energy, Solar Energy Res. List File, Golden, Colorado 80401, 1979
9. Chibata, I., Tosa, T., Kato, J., Wada, M.: Am. Soc. for Microbiology, Abstr. of ann. meeting, p. 202, Washington, D.C. 1979
10. Cysewski, G. R., Wilke, R. C.: Biotech. Bioeng. *8*, 1125 (1977)
11. Cysewski, G. R., Wilke, C. R.: Biotech. Bioeng. *9*, 1421 (1978)
12. De Carvalho, A. V., Jr., Milfont, W. N., Jr., Yang, V., Trinidade, S. C.: Energetics, economics and prospects of fuel alcohols in Brasil. Proceedings Vol. III, Volkswagenwerk, A.G. (ed.), Int. Symp. on Alcohol Fuel Techn. — Methanol and Ethanol, Wolfsburg, Nov. 21–23, 1977
13. Del Rosario, E. J., Lee, K. J., Rogers, P. L.: Biotech. Bioeng. *8*, 1477 (1979)

14. Dever, D. A.: The economic and social impacts of large scale utilization of renewable biomass energy. A Conference on Capturing the Sun through Bioconversion. Proceedings p. 707. Washington, D.C., March 10–12, 1976
15. Dunlap, C. E., Thompson, J., Chiang, L. C.: AIChE Symp. Ser. *72* (158), 58 (1976)
16. Egamberdiev, N. B., Jerusalimskii, N. E.: Mikrobiologie *37* (4), 686 (1968)
17. Engelbart, W., Dellweg, H.: Basic data on continuous alcoholic fermentation of sugar solutions and of mashes from starch containing raw materials. Proceedings Vol. III, Volkswagenwerk, A.G. (ed.), Int. Symp. on Alcohol Fuel Techn. — Methanol and Ethanol, Wolfsburg, Nov. 21–23, 1977
18. Ericsson, E. O.: Chem. Eng. Prog. *43* (4), 165 (1947)
19. Ghose, T. K.: Microbial technology in the provision of energy and chemicals from renewable resources. A Status Report on some areas, tasks and international coordination. Submitted to the Fermentation Commission, IUPAC, September 1978
20. Ghose, T. K., Tyagi, R. D.: Biotech. Bioeng. *8*, 1387 (1979)
21. Griffith, W. L., Compere, A. L.: Developments in Industrial Microbiology *17*, 241 (1976)
22. Hayes, R. D., Timbers, G. E.: Alcohol fuels from agriculture — A discussion paper. Engineering and Statistical Research Institute, Report 1–165, 1980
23. Nand, Krishna, Srikanta, S., Murthy, V. S.: J. Food Science and Techn. *14*, 80 (1977)
24. Lindeman, L. R., Rocchiccioli, C.: Biotech. Bioeng. *7*, 1107 (1979)
25. Mandels, M., Andreotti, R. E.: Process Biochemistry *5*, 6 (1978)
26. Margaritis, A., Wilke, C.: Biotech. Bioeng. *5*, 727 (1978)
27. Marion, L.: Chem. Eng. *86* (5), 78 (1979)
28. Meyers, G. S.: Ethanolic fermentation during enzymatic hydrolysis of cellulose. Proceedings: Pacific Chemical Engineering Congress, Biochemical Sources of Energy, p. 781, 1977
29. Nathan, R. A. (ed.): Fuels from sugar crops. Published by Technical Information Center, U.S. Dept. of Energy. USDOE Technical Information Center, Oak Ridge, Tenn., U.S.A. 1978
30. Paige, D., Boulton, R.: The production of ethanol from agricultural waste: An economic evaluation. California Univ., Davis: Final Report No. PB-292 552, 1979
31. Park, W. R.: The near term potential of biomass-based alcohol-gasoline transportation fuels. Proceedings The Second Ann. Symp. on Fuels from Biomass. June 20–22, Vol. 1, p. 77. Rensselaer Polytechnic Inst., Troy, N.Y. 12181, 1978
32. Ramalingham, A., Finn, R. K.: Biotech. Bioeng. *4*, 583 (1977)
33. Remirez, R.: Chem. Eng. *87* (6), 57 (1980)
34. Remirez, R., Grover, R., Marion, L.: Chem. Eng. *87* (5), 80 (1980)
35. Ressen, L., Strube, R.: Process Biochem. *11*, 21 (1978)
36. Scheller, W. A., Mohr, J. B.: Am. Chem. Soc. Div. of Fuel Chemistry, Preprints, Vol. 21, No. 2, p. 29, 1976
37. Scheller, W. A.: Nebraska Gasohol. Proceedings Vol. II, p. 2–1 (1), Volkswagenwerk, A.G. (ed.), Int. Symp. on Alcohol Fuel Techn. — Methanol and Ethanol, Wolfsburg, Nov. 21–23, 1977
38. Schruben, L. W.: Gasohol as an alternate fuel. Presented at the Kansas Agricultural Experiment Station Conference, Manhattan, U.S.A., Jan. 8, 1980
39. Sitton, O. C., Foutch, G. L., Book, N. L., Gaddy, J. L.: Chem. Eng. Prog. *75* (12), 52 (1979)
40. Sitton, O. C., Magruder, G. C., Book, N. L., Gaddy, J. L.: Comparison of immobilized cell reactor and CSTR for ethanol production. For presentation at Biotech. in Energy Production and Conservation Symp., Gatlinburg, Tennessee, Oct. 1979
41. Spano, C. A.: Enzymatic hydrolysis of cellulosic wastes to fermentable sugars for alcohol production. Chem. fuel from biomass, sewage, waste refuse, agricultural wastes, p. 325. Symp. sponsored by Inst. of Gas Techn., Chicago, Ill., Jan. 27–30. Orlando, Florida, U.S.A. 1976
42. Spano, L., Tassinari, T., Ryu, D. Y., Allen, A., Mandels, M.: Enzymatic hydrolysis of cellulose to fermentable sugar for production of ethanol. Presented at 29th Can. Chem. Eng. Conf., Sarnia, Ontario, Oct. 1979
43. SRI International, Merlo Park, Calif., U.S.A. under purchase order for U.S. Dept. of Energy, Fuel from biomass, Branch. Preliminary economic evaluation of a process for the production of fuel grade ethanol by enzymatic hydrolysis of an agricultural waste (Final Report, Jan. 1978)
44. Su, T. M.: Am. Inst. of Chem. Eng. *74*, 75 (1978)

45. Takagi, M., Abe, S., Suzuki, S., Emert, G. H., Yata, N.: Proceedings: Bioconversion of Cellulosic Substances into Energy, Chemicals and Microbial Protein. Ghose, T. K. (ed.), p. 551, New Delhi, Feb. 21–23, 1977
46. Tan, T. C., Lau, C. M.: J. Singapore Natl. Acad. Sci. *4* (3), 152 (1975)
47. Tatsumi, C., Ogaki, M., Tanabe, J.: British Patent *1*, 493, 480, 1977
48. Tyagi, R. D., Ghose, T. K.: In: Proceedings: Bioconversion of Cellulosic Substances into Energy, Chemicals and Microbial Protein. Ghose, T. K. (ed.), p. 585, New Delhi, Feb. 21–23, 1977
49. Watson, C. A.: Forest Products J. *3*, 25A (1959)
50. Wayman, M., Lora, J. H., Gulbinas, E.: Am. Chem. Soc. *90*, 183 (1979)
51. Weisz, P. B.: Population and energy, CHEMTECH, Vol. 10, No. 3, p. 131, 1980
52. Wilke, R. C., von Stockar, U., Yang, R. D.: Process design basis for enzymatic hydrolysis of newsprint: Energy, renewable resources and new foods. AIChE Symposium Series, p. 104, 1976
53. Rogers, P. L., Lee, K. J., Tribe, D. E.: Biotechnol. Lett. *1*, 165 (1979)
54. Lee, K. J., Tribe, D. E., Rogers, P. L.: Biotechnol. Lett. *1*, 421 (1979)
55. Lee, K. J., Skotnicki, M. L., Tribe, D. E., Rogers, P. L.: Biotechnol. Lett. *2*, 339 (1980)

Biomass Conversion Program in Brazil

José O. B. Carioca, Harbans L. Arora, Ahmad S. Khan
Nucleo de Fontes Não Convencionais de Energia da UFCe, Centro de Ciências — Camp us do Pici, Caixa Postal 935, 60.000 Fortaleza — Ceará Brazil

Brazil has based its economic development largely on imported petroleum and petroleum based technologies. The high hike in the price of this fossil fuel and uncertainties in its availability have forced the country to search for alternative fuels and chemicals for substitution of petroleum derivatives. The return to the traditional use of biomass as fuel, but employing more efficient processes and technologies specially for liquid fuels, has been taken as a challenge to combat petroleum crisis.

The paper describes and discusses briefly the technological state of art, the socio-economic and environmental impacts of a program of large biomass cultivation, conversion into biofuels and chemicals, and their use as petroleum derivatives substitutes.

1 Introduction

Most countries have been adversely affected by the petroleum crisis. The effect has been worse on those which had based their economic growth largely on imported petroleum. The cheap price and easy availability of this fossil fuel were responsible for the installation of an automotive transport system and petroleum based industries in Brazil.

The overnight trippling of petroleum prices in 1973 and their unpredictable steep hikes ever since as well as ucertainties in its obtention are bound to have grim consequences for imported petroleum dependent countries. For developing countries, the new situations are offering impediments to the pace of development. Brazil importing more than 80% of its petroleum needs had naturally to share the consequences of petroleum crisis.

Substitution of petroleum needs a multidisciplinary research and development of sources, processes and technologies for production and use of liquid fuels. For Brazil, the highly viable source for substitution of petroleum derivatives is biomass due to territorial vastness of the country and its localization in a tropical region extremely favorable for production of biomass. Furthermore, the country has been

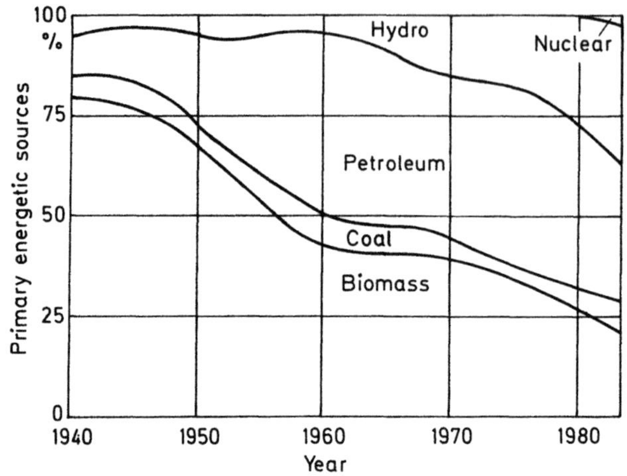

Fig. 1. Evolution of primary energetic sources participation in Brazil

traditionally utilizing biomass as a primary source of energy even during the period when the price of petroleum in the international market was low.

Wood has always played a very important role as a source of energy for food cooking in rural areas, in production of charcoal for steel industry and even as fuel for locomotives in the decades of the 40's and 50's.

There has been secular tradition in the sugar industry of Brazil for the use of bagasse as a primary source of energy. Moreover, the production of ethanol from sugar has always been an alternative whenever there was a fall in sugar prices in the international market.

However, the accelerated process of industrial development, mainly in the Southern region, starting from the decade of 1950, occurred in the automobile industry which made us strongly dependent on liquid fuels.

Figure 1 [12] shows the evolution of primary energetic sources in our country since 1940. It can be seen that in 1940, 77% of our basic energy consumption originated from biomass (wood, bagasse and charcoal), 7% from coal, 7% from hydropower and only 9% from petroleum. Between 1940 and 1980, the employment of biomass energy has been reduced to one third whereas that of petroleum increased 5-fold.

Furthermore, accompained by this development, there occurred an accelerated process of importation of associated technologies. Conjugation of these facts associated with elevated prices of petroleum have led the country to increasing external debts.

It is within this context that the biomass program for energetic and non-energetic aims presents itself as one of the best strategies for development of the country.

In spite of the traditional use of biofuels in Brazil, it was only in 1978 [2] that after the realization of "I Seminar about Energy from Biomass in the Northeast" that the basis for the establishment of a biomass program for energetic and non-energetic goals was delineated with a global outlook.

In the following sections, we shall discuss the technological and socio-economic aspects of the bioconversion program in Brazil.

2 Technological Aspects

Under the impact of steep elevation in the price of petroleum since 1973, the National Economic Development Council-CDE, an organ of the Presidency of the Republic, looked for the identification of indispensable elements for establishment of a National Alcohol Program — PROALCOOL in 1975, with the purpose of substitution of gasoline and petrochemicals.

Initially, this program aimed at production of alcohol in distilleries annexed to the sugar factories. In the second phase, autonomous distilleries based on sugar cane are being installed. In Table I the number of existing distilleries, as well as those to be implanted under PROALCOOL, is given.

Parallel to these efforts, the brazilian Government implanted — through its organ PETROBRAS and with the help of the National Institute of Technology INT — a

Table 1. Statewise distribution of existing and projected annexed and autonomous distilleries — July 1978

	Existing		Projected	
	Annexed	Autonomous	Annexed	Autonomous
North and Northeast	48	12	13	21
Amapá	—	—	—	—
Rondonia	—	—	—	—
Roraima	—	—	—	—
Acre	—	—	—	—
Amazonas	—	1	—	1
Pará	1	—	—	—
Maranhão	1	—	—	4
Piauí	1	—	—	3
Ceará	2	—	1	5
R.G. do Norte	2	1	—	1
Paraíba	5	4	—	2
Pernambuco	27	1	2	2
Alagoas	7	4	10	2
Sergipe	2	—	—	—
Bahia	—	1	—	1
Central/South Region	107	14	4	42
Minas Gerais	8	2	3	3
Espirito Santo	1	—	—	2
Rio de Janeiro	15	3	1	4
São Paulo	74	6	—	11
Paraná	3	2	—	11
Santa Catarina	3	—	—	2
R.G. do Sul	—	—	—	—
Mato Grosso	1	1	—	3
Goiás	2	—	—	6
Distrito Federal	—	—	—	—
Brazil	155	26	17	63

Source: Ministerio da Industria e Comercio — Secretaria de Tecnologia Industrial, Projeto de Referencia 1980

technology for the production of alcohol from cassava. A factory with the capacity of 60,000 $l\,d^{-1}$ was installed in the municipality of Curvelo in the State of Minas Gerais [19].

Another miniunit, also based on cassava, with a capacity of 12,000 $l\,d^{-1}$ was installed by our Nucleus in the municipality of Caucaia (State of Ceará) for research and development of improved technologies in the process of alcohol production from amilaceous materials such as cassava [7] and babassu mesocarp [6].

Within the same spirit and considering the necessity to utilize less noble lands for the production of energy (fuels), the Brazilian Government decided in 1979 [9] the implantation of a program for production of ethanol from wood. Main purpose of this program is to achieve more flexibility for PROALCOOL, making it independent of the fluctuations in the price of sugar in the international market.

In spite of the option made in favor of ethanol as gasoline substitute in Ottocycle motors, the Brazilian Government decided also to develop the technology for production of methanol from wood. To achieve this purpose, the Energetic Company of São Paulo — CESP, in collaboration with the Institute of Technological Research of São Paulo — IPT has already developed the basic engineering for production of 1,000 $t\,d^{-1}$ of this fuel from wood [3]. Presently, research is being conducted at the pilot plant level for the development of technology for gasification, which utilizes electrical energy in the process.

The technology for production of ethanol from mollase or sugar cane is according to our point of view, sufficiently old and therefore is in need for urgent improvements. At the I Seminar about Energy from Biomass in the Northeast [2], the possibility to utilize diffusors was indicated for substitution of traditional grinders for the purpose of increasing the quantity of extracted sugar. As emphasized by Pratas [15], our process for production of alcohol from sugar cane can be substantially improved through utilization of diffusion technology, increasing the yield from 66–77 $l\,t^{-1}$.

In relation to alcoholic fermentation, only recently a company — Usina Santa Adelaide of São Paulo is introducing the continous process, in spite of the tremendous efforts dedicated for more than two decades by Professor Walter Borzani of the University of São Paulo for development of this process in the country [4]. Much research work still needs to be done in the field of yeast which could work at elevated alcoholic concentrations or at temperatures higher than 35 °C.

The technology for distillation employed in the country was introduced in the decade of 1940 and remains unchanged until even today, in spite of the incredible technological development introduced in the distillation columns in petrochemical processes. With the use of bubble cap and direct injection of vapor, the country is spending approximately 100 million dollars per harvest due to problems caused by corrosion [13].

Only now with the installation of Caucaia plant, we are introducing a system of distillation with perforated trays and employing vapor in indirect form for the distillation of alcohols for surmounting the corrosion problem as well as for production of a reduced amount of stillage.

The enormous volume of stillage presently produced, about 50 billion l per year, is continuing to constitute a great threat to our ecology and consequently to PROALCOOL.

Four alternatives are being discussed for treatment of this enormous quantity of effluent from the distilleries. The first alternative, which is being mostly employed presently consists in its utilization "in natura" as fertilizer. With a proper dosage employed in an adequate manner, its use has shown an increase in agricultural productivity. The second alternative which is in early stages of implantation in Brazil is the process of concentration in multiple-stage evaporators for production of animal feed, fertilizer or fuel and fertilzer. The large majority of equipments produced in the country is based on foreign know-how, which justifies obviously the elevated price of these equipments. The third option consists in the anaerobic treatment of stillage for production of biogas and biofertilizers. This option appears to be becoming very useful for the country due to the possibility of improvements in the energetic balance of the distilleries in addition to production of fertilizers. A lot of effort is being dedicated for the development of this process in the country [8, 16, 18].

Finally, the last alternative seeks to produce single cell proteins. In spite of the fact that this process has been practiced in the country [1] it is still critical in the yield obtained. The research group of National Institute of Technology is undertaking much efforts for the improvement of the process.

In the same form as ethanol is an alternative fuel to gasoline, charcoal offers itself as an alternative primary material to mineral coal, considering the fact of our elevated external dependence, about 80% on this material.

Since Brazil does not possess high quality coal for steel industry, it turned to the use of charcoal and until 1946 all our steel industry was totally based on this renewable reducing agent. This fact made possible the development of steel technology based on charcoal. Presently, 50% of the total production of gusa in the country employs charcoal.

According to the National Steel Plan, the production of lingots has increased from 5.3×10^6 t in 1970 to 20×10^6 t in 1980 [11].

In order to have an idea as to the importance of wood for the brazilian steel industry we present in Table 2, some data about the production of gusa using

Table 2. Steel production, charcoal consumption and estimated reforestation area

Year	Steel production (gusa) (10^6 t)	Charcoal consumption (10^6 t)	Charcoal production from reforestation (10^6 t)	Eestimate forest plantation area (ha)
1975	3.575	13.059	0.759	58,669
1976	4.825	17.338	1.152	90,117
1977	5.011	17.677	1.660	92,879
1978	5.641	19.429	2.213	98,656
1979	5.935	20.448	2.587	101,421
1980	6.108	21.069	2.573	101,963
1981	6.112	21.107	3.265	103,485
1982	6.182	21.051	4.896	97,983
1983	6.186	21.063	5.834	92,725
1984	6.151	20.930	6.857	93,838
1985	6.195	21.093	10.300	97,826

Source: adapted from CONSIDER, Ministerio de Indústria e Comercio

charcoal, its national consumption, its production based on reforestation program as well as an estimate of plantation necessary to give support to the National Steel Program.

An analysis of this Table reveals that presently an enormous amount of charcoal produced in the country is obtained from native forests, which represents a big risk to our ecology.

In relation to the technology for production of charcoal, one notes that there is dominance of traditional processes. However, the present and future situations demand the development of modern technologies based on continuous processes, with utilization of volatile materials and with larger charcoal yield per ton of material employed.

Presently, the Foundation of Technological Center of Minas Gerais — CETEC located in Belo Horizonte — Minas Gerais, the largest gusa producing State based on charcoal is involved in an excellent program for development of this technology.

We believe that the development of this area will depend very much on corresponding improvement of wood hydrolysis (acid or enzymatic) presently in the phase of development in the country. This route would allow the integral use of wood, to the benefits of the steel industry (lignin), the energy sector or more specifically liquid fuels (ethanol) and finally the food sector (SCP).

These programs have been implanted as an action of some isolated industrial sectors due to the steep rise in the price of petroleum and increasing dependence of Brazil on foreign energy sources (petroleum and coal) of the order of 80%. These programs are a part of the Brazilian Energetic plan, whose projects envisage a reduction of our present external petroleum dependence from 80% to about 30% in 1985, respectively.

Table 3 represents the main objectives of this model.

Considering the fact that there is an urgent necessity to search for options for substitution of diesel oil, the Brazilian Government is presently involved in a program of alternative fuels starting from vegetal oils — PRODIESEL.

With the progress in gasoline substitution by ethanol, diesel by vegetal oils, and industrial oils by coal, the country can largely reduce its petroleum imports.

Table 3. Brazilian energetic model

Projections (1985)	Energy (barrels equivalent petroleum/day) (10^3)
Demand	1,700
Conservation	200
Petroleum (imported)	500
Petroleum (national)	500
Coal	170
Charcoal	120
Shale	25
Other sources	15
Alcohol	170

Source: Ministerio das Minas e Energia: Modelo Energetico Brasileiro. Brasilia, Novembro 1979

In technological terms, two processes represent potentialities, which can transform vegetal oils, generally triglyceroids, as alternative fuels to diesel oil [10]. These processes are transesterification and therm-catalytic cracking, both being developed in the country [14,17].

There exists a big potential for production of biogas and biofertilizers through anaerobic fermentation of animal residues, considering the fact that there are more than 120 million cattles in the country: Specifically for the rural areas where about 40% of the population lives, and where more than 90% of rural farms are not yet electrified, anaerobic degradation of animal residues could be a very pragmatic approach for energization of rural areas. It is only within the last few years that Federal and State Governments are stimulating installation of biodigestors of mainly Indian and Chinese types. This technology is propagated through demonstration units and provision of low interest credits. Success of these programs is leading to the definition and establishment of a Government policy for rural energization.

For better utilization of agricultural residues, the traditional gasogene technology has been developed at the industrial level and is expanding throughout the country. Actually, a number of units are functioning in farms and industries for power generation.

Recently, the Ministry of Mines and Energy and the Ministry of Agriculture have established a joint program mapping the zones destined to the production of energy and food in the country. This study will also provide the acquisition of information on the real potentiality of primary materials for food and energy.

Studies conducted by various groups in Brazil have demonstrated the importance of decentralized integrated systems which make adequate use of subproducts and residues from energy producing systems for improvement of energetic and economic balance. Such systems would allow the development of agriculture and of agroindusstries for the production of food, feed and fuels.

Regarding substitution of petrochemicals the country is developing appropriate routes to produce chemicals such as ethylene, butadiene and acetyldehyde etc, from ethanol. It should be noted that out of 10.7 billion l of alcohol projected to be produced in 1985, about 15% would be utilized for alcoholchemicals.

3 Socio-Economic and Environmental Aspects

The biomass conversion program has helped to increase the gross national product. For example, in 1980, GNP of the country increased by 8.5%, highest among all the non-oil exporting countries. However, the increase in GNP and high per capita income are not necessarily best indicators of development of a country. In fact, the expansion of the economy must be accompanied by better distribution of income. However, the PROALCOOL is leading to the accumulation of wealth in the hands of a few agro-industrialists due to the low interest credit scheme.

The PROALCOOL is helping to reduce country's vulnerability caused by the uncertainty associated with foreign petroleum supply. For example, in 1980, it brought a reduction of about 10,000 $m^3\ d^{-1}$ in imported oil.

It is also estimated that only PROALCOOL will require an investment of $ 5 billion for installation of new distilleries up to 1985 and generate savings of much higher value in hard currencies, consequently reducing the pressure on National Balance of Payment.

The Biomass Conversion Program has raised public awareness of the important role that biomass can play for the future development of the country. It has also provided an opportunity to scientists and technologists to develop new and appropriate national technologies to reduce country's dependence on imported ones. It should be remarked here that Brazil is already spending a substantial amount of foreign exchange for importing technologies for various sectors of the economy.

In fact all forms of biomass suitable for energy can have non-energy uses as well, and bioenergy production will compete with other uses for the same land base. In most of the developing countries like Brazil, agriculture provides food in form of corn, rice, wheat, beans etc. for local population and earns foreign exchange through export of cash crops such as coffee, cotton, soyabeans etc. It is expected that, in 1985, 3 million ha of fertile land will be needed for sugar cane cultivation to reach the target of 10.7 billion l of ethanol production. It is estimated that these 3 million ha can produce 5.1 million t of corn or 1.5 million t of beans or 4.5 million t of rice or various combinations of these crops. It may be remarked here that if the Government also decides to substitute 2.5 million l of diesel oil by vegetable oils, then additional 4 to 5 million ha of fertile land will be needed. The transfer of land from food crops to biomass for energy production will further aggravate the existing deficit food situation of the country [5].

Furthermore, the expansion of a big agriculture program would necessitate an extensive use of fertilizers, which is a serious problem to be tackled since at present the country imports large quantity of fertilizers.

The program for production of alcohol and diesel substitutes based on biomass will create about one million jobs during 1980–85 which is very important for a country having a high population growth rate. Unfortunately, 80–85% of these jobs will be available in the South and Central South regions of the country which are already rich in agriculture and also have most of the industry of the country. This situation will further stimulate the existing migratory trend from North and Northeast to these regions and thereby create social problems.

Out of the 228 distilleries projects approved by the end of 1979, with the exception of one, all are based on sugar cane as feedstock. The increased commercialized farming of sugar cane necessary for fermentation alcohol, particularly in the State of São Paulo, is leading to dislocation of small land owners and rural workers to cities, augmenting urban employment problems.

Biomass feedstocks (sugar cane, soyabeans, nuts, babassu etc.) may be diverted for nonenergy uses (e.g., sugar, vegetable oils for human consumption) that may, at times, have a greater economic value. Recent big increase in the price of sugar in the international market is bound to influence the supply of sugar cane for ethanol production. Adverse weather conditions also can interrupt harvesting or reduce biomass productivity and thus create imbalance between the biomass quantities produced and consumption needs. Moreover, if any of these factors should cause biomass supply problems, high transportation costs for supply of primary feedstock from other regions may make such problems difficult to solve. It may be mentioned here that some of the capital intensive high capacity distilleries in the country are already running below full capacity. These aspects stress the need for installation of small and medium sized systems based on diversified primary feedstocks.

Biomass in Brazil although has the potential of an energy source may create serious environmental problems. The feedstocks have to be harvested properly and the conversion technologies adequately chosen. For example, the production of 10.7 billion l of alcohol, in 1985, will be producing 130–155 billion l of stillage creating a pollution potential about twice that of the present population. Stillage is being released in the rivers and fields creating air and water pollution. There is an urgent need to establish a national program for the development of efficient biogas, biofertilizers or animal feed bioprocesses.

4 Present Status and Future Trends

The program of substitution of petroleum derivatives by biofuels is in a fairly advanced stage in Brazil. The National Alcohol Program has quite successfully met its present target. To reach the goal of 1985 for production of 10.7 billion l of alcohol, more than two million ha of additional fertile land must be brought under sugar cane cultivation. The competition with food and feed crops and necessity of large quantities of fertilizers represent serious problems. The diversification of primary feedstocks by cultivation of amilaceous (cassava, babassu, etc.) and cellulosic (wood) materials using marginal lands is very essential.

The technology for conversion of vegetal oils to diesel oil substitute is in an advanced stage of development and offers good future prospects.

The conversion of wood into ethanol, charcoal and single cell proteins, through the processes of wood hydrolysis, is being seriously studied.

The program for production of gaseous fuels from animal and biomass residues has been recently undertaken and is being gradually expanded in rural areas.

The formulation of a broad-based policy for development of agriculture and agro-industries compatible with food, feed, fuel and chemicals production, integrated with adequate use of subproducts and residues is being seriously considered to serve as a guideline for future developments.

5 References

1. Ann. Seminario Intern. sobre Tratamento do Vinhoto. Inst. Nac. Tecnologia, INT, Rio de Janeiro, Aug. 1976
2. Ann. I Seminário sobre Energia de Biomassa no Nordeste. Fortaleza — Aug. 1978
3. Antonio, J. A., Zagatto, G.: Engenharia Química. *3*, 44 (1980)
4. Borzani, W.: Fermentação Contínua, IV Simpósio Nacional sobre Fermentação. Recife 1980
5. Brown, L. R.: Food or Fuel: New Competition for the World's Crop Land. World Watch Paper 1980
6. Carioca, J. O. B., Soares, J. B.: Babaçú: Uma Fonte Não Convencional de Energia, "I Simpósio sobre Produção de Álcool", B.N.B. Fortaleza, Brazil 1976
7. Carioca, J. O. B., Arora, H. L., Khan, A. S.: Technological and Socio-Economic Aspects of Cassava-Based Autonomous Minidistilleries in Brazil. Submitted for Publication in "Biomass" (1980)
8. Carioca, J. O. B., Arora, H. L., Carvalho, A.: Biodigestor Economico para Aproveitamento do Vinhoto, "Congresso Brasileiro de Engenharia Química" 1980

9. COALBRA: Coque e Álcool da Madeira. Technical Report. 1980
10. IV Seminario Internacional sobre Tecnologia do Álcool como Combustível — Guarujá, São Paulo — Oct. 1980
11. Faria, L. de M.: Metalurgia. *30*, 237 (1974)
12. Goldemberg, J.: Energy Options and Current Outlook. Reprint. Institute of Physics, University of São Paulo 1977
13. Michaelles, M.: Jornal do Brasil. I Caderno. Sept. 1980
14. Núcleo de Fontes Não Convencionais de Energia: Internal report 1980
15. Pratas, M. C.: Engenharia Química *4*, 20 (1980)
16. Pratas, M. C.: Personnal Communication
17. PROERG: Techn. Bull., Fortaleza, Brazil 1980
18. Ribeiro, C. C., Branco, J. R. C., Lacaz, P. A.: Petro & Química. Abril, 55–60 (1979)
19. Thorio, B. S.: Possibilidade de Produção de Álcool a partir da Mandioca. In: I Simposio sôbre produção de Álcool no Nordeste, Fortaleza, Brazil 1976

Biomass Conversion Program in Finland

Matti Linko
Technical Research Centre of Finland, Biotechnical Laboratory
Tietotie 2 SF-02150 Espoo 15 Finland

The production and use of cellulolytic enzymes has been studied in Finland for several years. Major areas of interest include enzyme biochemistry, development of activity determination methods, genetic development of the cellulase producing organism *Trichoderma reesei*, studies on enzyme production in laboratory scale and in a computercoupled pilot fermentor, enzymatic hydrolysis of cellulose and hemicellulose, production of ethanol and single-cell protein, treatment of silage with cellulases and stimulation of malting by cellulases. The research is ultimately aimed at full scale process applications.

1 Introduction

A new potential of cellulosic materials for the production of energy, protein and chemicals has been recognized in many countries, including Finland, a land covered with large forests. The ultimate goals of the research programs related to the utilization of biomass are full scale processes for production of the necessary enzymes and for hydrolysis of biomass to sugars which can be used e.g. for production of ethanol or single-cell protein. However, basic information in many fields is needed before these goals can be attained. Therefore, the research programs also include biochemistry, analytical chemistry, systems science, microbial genetics etc. This article reviews the research activities at the Biotechnical Laboratory of the Technical Research Centre of Finland and at the Helsinki University of Technology.

2 Cellulase Activity Determinations

Specific methods for determination of the different types of cellulolytic enzymes are needed [1]. In this context, the synergistic action of the enzymes presents a difficult problem. Methods measuring the solubilizing effect give useful information for practical applications [2]. A method based on the release of color from a dyed microcrystalline cellulose substrate was developed [3,4,5,6,7]. The method has also been modified for automatic analysis [8]. An autoanalyzer connected to the computer controlling a bioprocess opens new possibilities for process control strategies [9,10,11,12,13]. The production of cellulase can be seen on the graphic display of the computer (Figs. 1 and 2).

An immunoelectrophoretic detection method for cellulases was developed for biochemical studies [14,15].

3 Production of Enzymes

3.1 Microbial Genetics

Trichoderma reesei strain QM 9414 was used as the basis for further genetic development. The mutagenic treatments used included γ-irradiation as well as treat-

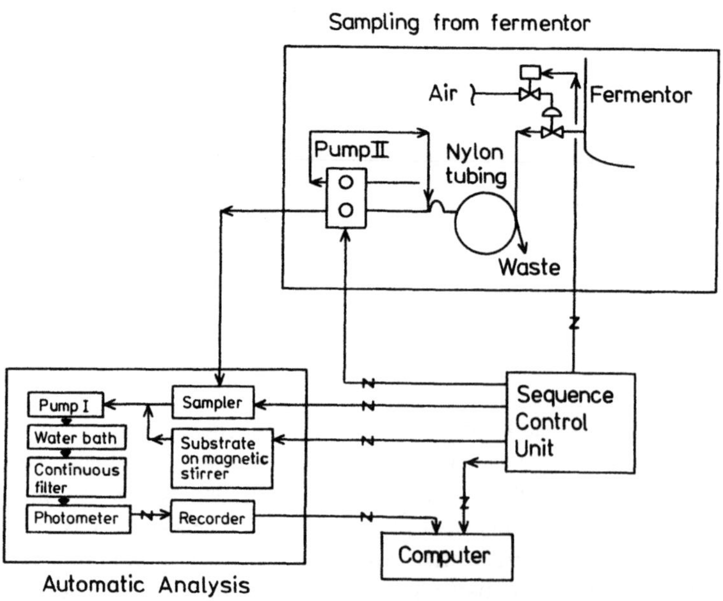

Fig. 1. Automated cellulase analysis system connected to the computer of a pilot reactor

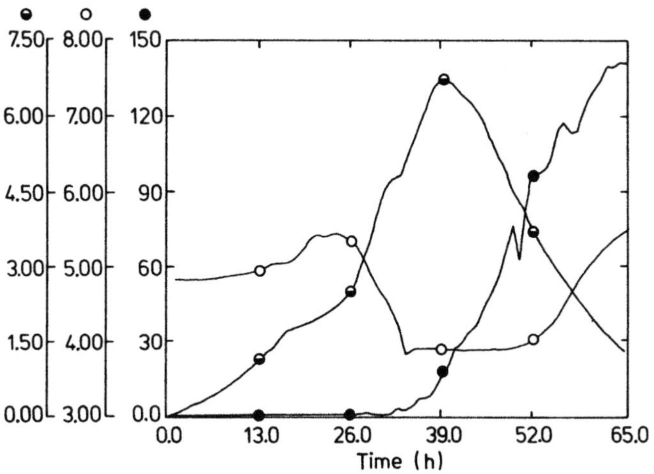

Fig. 2. Production of cellulase by *Trichoderma reesei*. Hard copy from the graphic display of the computer. Cellulase activity (●——●), pH (O——O), evolution of CO_2 (O——O)

ments with alkylating chemicals [16, 17, 18, 19, 20, 21]. The total number of colonies tested was about 30,000. The mutation steps have been presented in Fig. 3. The new mutants were clearly superior to QM 9414 as cellulase producers.

3.2 Cellulases

Many organisms, especially fungi, produce one or several of the enzymes needed for hydrolysis of cellulose and hemicellulose, but it is not very likely that any single organism would be the most efficient producer of all these enzymes [22]. *Aspergillus* molds are good producers of some enzymes [23], but *Trichoderma reesei* is the best known producer of extracellular cellulases capable of hydrolyzing native cellulose. The effects of the composition and concentration of cultivation media and of cultivation conditions have been studied using *T. reesei* [24, 25, 26]. Ground sulfite cellulose at a concentration of 3–5% was found to give high activities in fermentor cultivations. Cellulose can partly be replaced by lactose or whey. The presence of organic nitrogen is essential for efficient enzyme production. Distiller's spent grain together with cell mass from previous cultivations gave rise to high activities.

3.3 β-Glucosidase

When hydrolyzing cellulose with an enzyme preparation produced by *T. reesei* the final stage, the breakdown of cellobiose to glucose, is often rate limiting. There are two reasons for this: *T. reesei* does not produce very high levels of β-glucosidase and the enzyme is subject to end product inhibition [27].

Some new *T. reesei* mutants produced several fold β-glucosidase activities compared to the strain QM 9414 [20, 21], but *Aspergillus phoenicis* and *Aspergillus niger* were found to be superior as β-glucosidase producers [27, 28]. A typical comparison has been

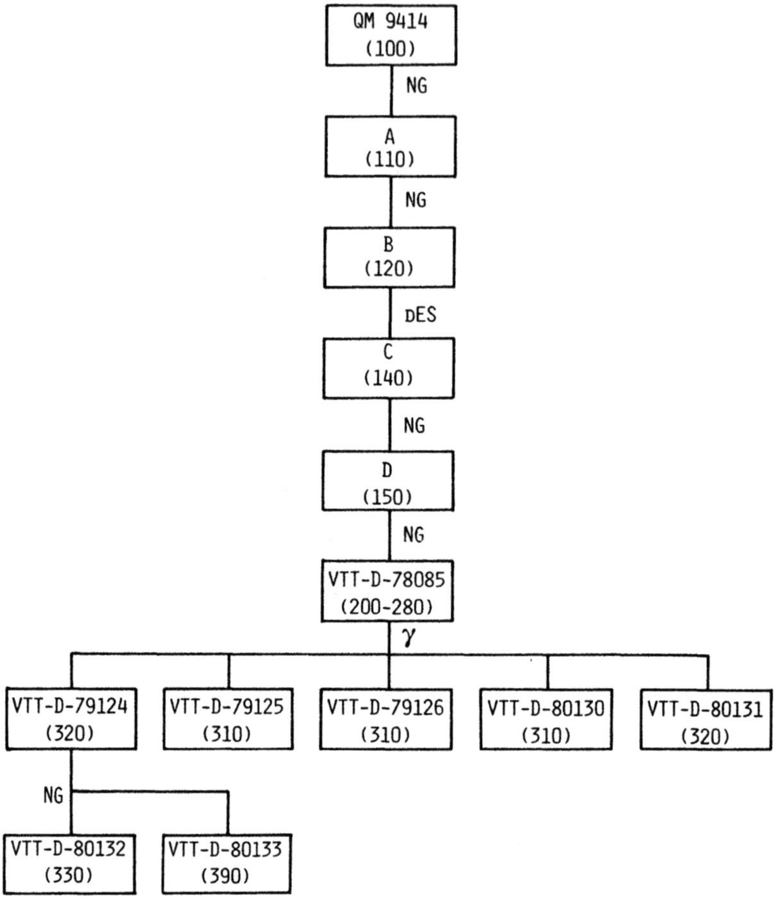

Fig. 3. Stages in the development of high-cellulase *T. reesei* mutant strains. Figures in parentheses represent production of cellulase as percentages of the production by strain QM 9414. Mutagenic agents: NG = N-methyl-N′-nitro-N-nitrosoguanidine; dES = diethyl sulphate; γ = gamma irradiation Co^{60})

presented in Fig. 4. The production of β-glucosidase by *A. niger* was found to be constitutive: good production was observed using starch, starch syrup, glucose or glycerol as sole carbon source. This enzyme is not subject to end product inhibition at the glucose concentrations involved in the hydrolysis of cellulose.

3.4 Xylanases

In many potential technical applications both cellulose and hemicellulose should be hydrolyzed. The hemicellulose of hardwoods is mainly of xylan type. *Trichoderma reesei, Aspergillus foetidus* and *Bacillus subtilis* were tested for their xylanase production capacity [25, 29, 30]. *T. reesei* produced the highest activity, but the time needed for maximal enzyme production was about 7 days, whereas *A. foetidus*

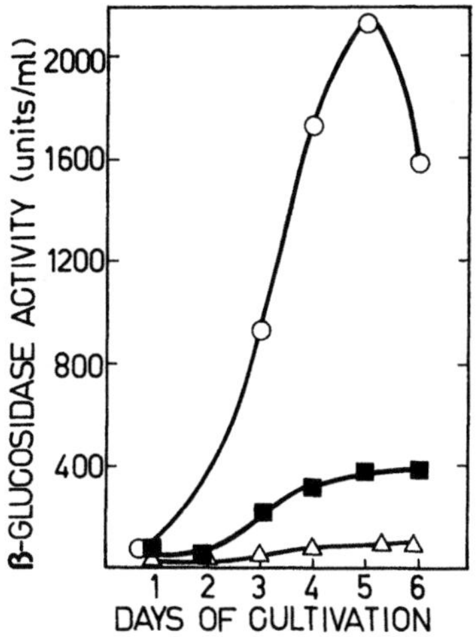

Fig. 4. Production of β-glucosidase by two *T. reesei* strains (△——△, ■——■) and by *Aspergillus phoenicis* (○——○)

produced relatively high activities in 3 days (Fig. 5). The latter organism also produced the highest β-xylosidase activity. The xylanase complex of *B. subtilis* proved to be rather ineffective in prolonged hydrolyses of xylan.

3.5 Enzyme Production Technology

For the development of industrial enzyme production processes knowledge of microbiology, microbial genetics, biochemistry, biotechnology, chemical engineering and process control engineering must be combined into a working unity [31]. Problems

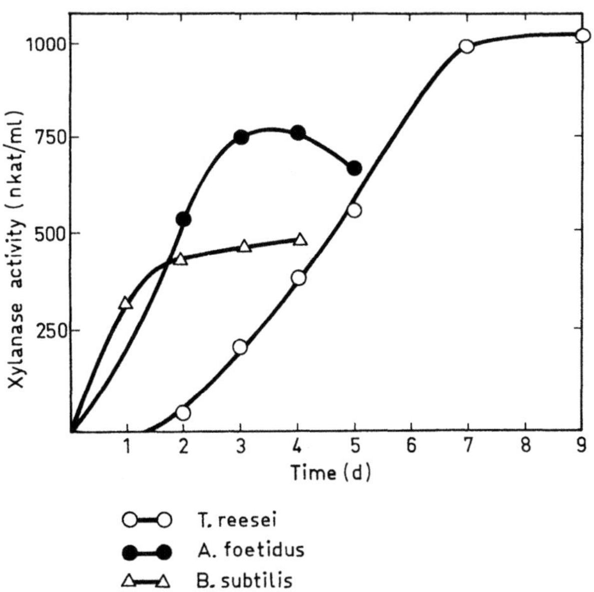

Fig. 5. Production of xylanase by *Aspergillus foetidus* (●——●), *Trichoderma reesei* (○——○) and *Bacillus subtilis* (△——△)

not fully appreciated in the laboratory may arise at the pilot plant or in industrial production. Therefore, a highly instrumented relatively large pilot bioreactor is valuable for process development [32]. Mathematical models for identifying and control of bioprocess are useful tools in improving the efficiency and productivity of a computer-coupled reactor [33, 34].

Continuous processes are often advantageous in biotechnology. Similar advantages to those obtained in a truly continuous process, such as elimination of dead-time between subsequent batches, may also be achieved in a semi-continuous process. It was possible to sustain semi-continuous cultivation of *Trichoderma reesei* with good production of cellulolytic enzymes for periods of several weeks [35].

4 Biochemistry of Cellulolytic Enzymes

Detailed biochemical information is necessary as a basis for process development. The formation of disaccharides acting as inducers for the *Trichoderma reesei* enzyme complex was investigated [36]. It was found that sophorose, laminaribiose and gentiobiose are formed both during the hydrolysis of cellulose by *T. reesei* culture filtrate and during incubation of cellobiose with a cell free extract of *T. reesei.*

For studies on the regulatory system of cellulases a method was developed for the production of cellulases using the soluble inducer cellobiose as the only carbon source [37]. The production was based on the continuous feeding of cellobiose to a batch culture. With a suitable feeding program the repression and inactivation by cellobiose could be avoided. Cellulase production was at least as high as that obtained with cellulose as carbon source.

The β-glucosidase enzymes produced by *Trichoderma reesei* and *Aspergillus niger* were purified and compared [38]. End product inhibition by glucose was much stronger for the *T. reesei* enzyme (K_i = 0.7 mM) than for the *A. niger* enzyme (K_i = 2.1 mM). It is thus advantageous to add *A. niger* β-glucosidase to *T. reesei* cellulase for the efficient hydrolysis of cellulose. This was verified by hydrolysis experiments with purified enzymes.

5 Applications Based on Hydrolysis of Cellulosic Materials

The economical feasibility of processes based on enzymatic hydrolysis of cellulosic materials is uncertain so far, but the potential of these processes encourages further development work [39, 40, 41].

5.1 Hydrolysis of Cellulose

Pretreatment is the major problem in enzymatic hydrolysis of cellulose, as efficient pretreatment methods tend to be expensive. Therefore, readily hydrolyzable materials such as the wastes from furfural or xylitol production processes are especially interesting. In these processes the pentosan fraction is used for the main product and the cellulose is simultaneously modified to some extent [42, 43, 44]. The cellulose of furfural process waste was almost completely hydrolyzed in one or two days with a *T. reesei* enzyme preparation.

5.2 Hydrolysis of Hemicellulose

The sugars originating from hemicellulose are typical constituents of sulfite waste liquor. Due to the relatively mild conditions during bisulfite pulping most of the dissolved hemicelluloses are only partially hydrolyzed. The resulting waste liquor can thus not be used for the production of single-cell protein. Both soluble and immobilized enzymes were tested for hydrolysis of the oligosaccharides in this type of waste liquor[45]. The hemicellulase enzyme was immobilized onto a phenol-formaldehyde resin using adsorption and cross-linking. Almost complete hydrolysis was achieved in a plug-flow reactor using a flow rate of one bed volume per hour at 40 °C, pH 4.5. The half-life of enzyme activity was about 30 days.

5.3 Production of Ethanol and Single-Cell Protein

For production of ethanol a simultaneous process for enzymatic hydrolysis of cellulose and ethanol fermentation is advantageous compared with a two-step process[46]. Continuous removal of glucose from the reaction mixture through the sugar conversion to ethanol prevents inhibition of the hydrolysis.

An alternative to yeasts for ethanol production is the bacterium *Zymomonas mobilis*. This organism has a faster specific rate of alcohol production and is capable of growing in higher glucose and ethanol concentrations[46]. The *Saccharomyces* yeasts and the *Zymomonas* bacteria can only convert hexoses to ethanol. However, some *Fusarium* molds are able to produce ethanol from both hexoses and pentoses, although this process is relatively slow[47]. A separate pentose conversion may be advantageous in order to increase the total ethanol yield and to improve the overall economics of the process.

Hydrolyzates of cellulosic materials are also suitable for production of single-cell protein[48,49,50]. Many organisms, such as *Candida utilis* and *Paecilomyces variotii*, utilize both hexoses and pentoses for their growth.

An interesting raw material for production of SCP or ethanol is peat[50]. The use of peat on a world-wide basis is about 200 million tons per year, which is only 0.08% of the total peat resources, estimated at 260 billion tons. These resources are particularly abundant in the Soviet Union, Canada and Finland. The cellulose and hemicellulose content of surface layers of peat is relatively high.

5.4 Treatment of Silage with Cellulases

Cellulases can be used to increase the digestibility and the sugar content of silages[48,51,52]. In Finland silage is produced mainly by the AIV method, in which the pH is lowered by acid solutions. The AIV solution most commonly used contains 80% formic acid and 2% phosphoric acid. In good silages made by the AIV method the sugar content is above 2%, the lactic acid content below 1% and the ammonia content below 0.5 $g\ l^{-1}$. The rumen microbes need energy to synthesize protein from ammonia and other soluble nitrogen compounds in feed. In practice it may be that the silages are deficient in energy rather than in nitrogen. Therefore, the production of more sugar in silage by cellulolytic enzymes may be advantageous.

5.5 Stimulation of Malting with Cellulases

A well-known stimulator of the malting process is gibberellic acid. However, the penetration of gibberellic acid into the grain is a limiting factor. Therefore, addition of microbial enzymes capable of breaking cellular structures may stimulate the malting process. A combination of gibberellic acid and cellulolytic enzymes holds promise for practical malting[53,54]. The viscosity of wort is decreased when cellulases are used in malting. This may be of advantage in the brewing process, because high viscosity often creates technical problems.

6 Present Status in Finland

The research on production and use of cellulolytic enzymes is continuing. Industrial production of cellulases may already be economically feasible with the new mutant strains of *Trichoderma reesei*. Further improvement may be achieved through the recombinant DNA technique. Some uses for cellulases are already envisaged at the present production cost, such as stimulation of the malting process and treatment of silage. Production of ethanol from cellulosic materials is probably not very far in the future.

7 Acknowledgement

The biomass conversion programmes in Finland have been supported by The Academy of Finland, The Ministry of Trade and Industry, The Finnish National Fund for Research and Development (SITRA), The State Fuel Centre, The State Alcohol Monopoly (ALKO) and Metsäliiton Teollisuus Oy.

8 References

1. Enari, T-M., Nybergh, P.: Kemia-Kemi *6*, 301 (1979)
2. Linko, M.: 8th North West Eur. Microbiol. Group Meet. Abstr., p. 44, Helsinki 1976
3. Leisola, M., Linko, M.: Symp. Enzymatic hydrolysis of cellulose (eds.) Bailey, M., Enari, T-M., Linko, M., p. 297, Helsinki, SITRA 1975
4. Leisola, M., Linko, M.: Anal. Biochem. *70*, 592 (1976)
5. Leisola, M., Linko, M.: Finnish Chem. Lett., p. 172 (1977)
6. Leisola, M., Linko, M.: Determination of cellulases with dyed substrates. Valtion teknillinen tutkimuskeskus, biotekniikan laboratorio, tiedonanto *13*, Helsinki 1976
7. Leisola, M., Linko, M.: 5th Int. Fermentation Symp. Abstr., p. 445, Berlin 1976
8. Leisola, M., Kauppinen, V.: Biotech. Bioeng. *20*, 837 (1978)
9. Leisola, M., Karvonen, E.: Symp. Bioconversion in Food Technology (ed.) Linko, P., p. 178, Espoo, Technical Research Centre of Finland 1978

10. Leisola, M. et al.: Enz. Microbial Techn. *1*, 117 (1979)
11. Karvonen, E., Leisola, M., Virkkunen, J.: Kemia-Kemi *6*, 309 (1979)
12. Karvonen, E., Meskanen, A.: Dechema-Kurs, Meß- und Regeltechnische Probleme in Bioreaktoren, p. 139, Zürich 1979
13. Karvonen, E., Ojamo, H., Linko, M.: 11 Farmacodegli anni '80, Montecatini Terme 1980
14. Niku-Paavola, M-L., Raunio, V., Nummi, M.: 27th IUPAC Congr. Abstr., p. 573, Helsinki 1979
15. Nummi, M. et al.: FEBS Lett. *113*, 164 (1980)
16. Nevalainen, H., Palva, T.: 8th North West Eur. Microbiol. Group Meet. Abstr., p. 114, Helsinki 1976
17. Nevalainen, H., Palva, T.: Appl. Environ. Microbiol. *35*, 11 (1978)
18. Palva, T., Nevalainen, H.: 2nd Nat. Meet. Biophysics and Biotechnology in Finland Proc., p. 93, Espoo 1976
19. Nevalainen, H., Palva, T., Bailey, M.: 27th IUPAC Congr. Abstr., p. 571, Helsinki 1979
20. Nevalainen, H., Palva, T., Bailey, M.: Enz. Microbial Techn. *2*, 59 (1980)
21. Bailey, M., Nevalainen, H.: Enz. Microbial Techn. *3*, 153 (1981)
22. Enari, T-M., Markkanen, P.: Production of cellulolytic enzymes by fungi. In: Advances in Biochemical Engineering. Vol. 5 (eds.) Ghose, T. K., Fiechter, A., Blakebrough, N., p. 1, Berlin, Springer 1977
23. Enari, T-M., Markkanen, P., Korhonen, E.: Symp. Enzymatic hydrolysis of cellulose (eds.) Bailey, M., Enari, T-M., Linko, M., p. 171, Helsinki, SITRA 1975
24. Markkanen, P., Bailey, M.: 5th Int. Fermentation Symp. Abstr., p. 444, Berlin 1976
25. Linko, M. et al.: Proc. Int. Symp. Bioconversion (ed.) Ghose, T. K., p. 329, New Delhi, Indian Institute of Technology 1977
26. Markkanen, P., Bailey, M., Enari, T-M.: Symp. Bioconversion in Food Technology (ed.) Linko, P., p. 111, Espoo, Technical Research Centre of Finland 1978
27. Bailey, M. et al.: 1st Eur. Congr. Biotechnology, Part 2. Poster Papers, p. 313, Interlaken 1978
28. Bailey, M., Nybergh, P.: 27th IUPAC Congr. Abstr., p. 544, Helsinki 1979
29. Viikari, L., Linko, M., Enari, T-M.: 1st Eur. Congr. Biotechnology, Part 1. Discussion Papers, p. 3/147, Interlaken 1978
30. Klemola, M., Viikari, L.: Kemia-Kemi *5*, 12 (1978)
31. Linko, M.: 2nd Nat. Meet. Biophysics and Biotechnology in Finland. Proc., p. 84, Espoo 1976
32. Virkkunen, J., Linko, M.: Application of systems science to biotechnical processes. Valtion teknillinen tutkimuskeskus, biotekniikan laboratorio, tiedonanto 20, Helsinki 1979
33. Nihtilä, M., Virkkunen, J.: Symp. Bioconversion in Food Technology (ed.) Linko, P., p. 140, Espoo, Technical Research Centre of Finland 1978
34. Virkkunen, J.: 27th IUPAC Congr. Abstr., p. 530, Helsinki 1979
35. Linko, M., Bailey, M., Markkanen, P.: 4th FEMS Symp. Abstr. B40, Vienna 1977
36. Vaheri, M., Leisola, M., Kauppinen, V.: Biotech. Lett. *1*, 41 (1979)
37. Vaheri, M., Vaheri, M., Kauppinen, V.: Eur. J. Appl. Microbiol. Biotech. *8*, 73 (1979)
38. Enari, T-M.: Proc. 2nd Int. Symp. Bioconversion (ed.) Ghose, T-K., New Delhi, Indian Institute of Technology 1980 (in press)
39. Linko, M.: Kemia-Kemi *2*, 602 (1975)
40. Linko, M.: An evaluation of enzymatic hydrolysis of cellulosic materials. In: Advances in Biochemical Engineering. Vol. 5 (eds.) Ghose, T. K., Fiechter, A., Blakebrough, N., p. 25, Berlin, Springer 1977
41. Linko, M.: Dechema-Monographien *83*, 209 (1978)
42. Markkanen, P., Eklund, E.: Symp. Enzymatic hydrolysis of cellulose (eds.) Bailey, M., Enari, T-M., Linko, M., p. 337, Helsinki, SITRA 1975
43. Mustranta, A., Nybergh, P., Hatakka, A.: 2nd Nat. Meet. Biophysics and Biotechnology in Finland. Proc., p. 96, Espoo 1976
44. Markkanen, P., Linko, M., Nybergh, P.: AIChE Symp. Ser. *74*, Nr. 172, p. 89 (1978)
45. Weckström, L., Leisola, M.: 6th Int. Fermentation Symp. Abstr., p. 89, London, Ontario 1980
46. Viikari, L., Nybergh, P., Linko, M.: 6th Int. Fermentation Symp. Abstr., p. 80, London, Ontario 1980
47. Viikari, L., Suihko, M-L., Enari, T-M.: 2nd Eur. Congr. Biotechnology, Eastbourne 1981

48. Linko, M.: Food and Agricultural Organization of the United Nations. New Feed Resources. Proc. technical consultation, Rome 1976
49. Hatakka, A. et al.: 2nd Nat. Meet. Biophysics and Biotechnology in Finland, Proc., p. 99, Espoo 1976
50. Mustranta, A. et al.: COST Workshop. Production and Feeding of Single Cell Protein. Abstr., p. 40, Jülich 1979
51. Vaisto, T. et al.: J. Sci. Agric. Soc. Finland *50*, 392 (1978)
52. Vaisto, T. et al.: Karjatalous *54*, Nr. 10, p. 26 (1978)
53. Home, S., Lehtomäki, I., Linko, M.: 27th IUPAC Congr. Abstr., p. 574, Helsinki 1979
54. Home, S.: Mallas ja olut, p. 21 (1980)

Biomass Conversion Program of West Germany

H. Sahm
Institut für Biotechnologie der Kernforschungsanlage Jülich, D-5170 Jülich, West Germany

Due to concern about the cost and availability of petroleum and natural gas, which are the main sources of energy (about 70%) and chemical feedstocks in West Germany, a great interest is increasing in research projects to discover new sources of energy and raw materials. Apart from nuclear and solar energy, hopes have also been placed in some biotechnological processes which utilize the so-called renewable resources. Several research and development projects are, therefore, supported by the Ministry of Research and Technology (BMFT) for the study of those microbial processes in more detail which are involved in the conversion of biomass — either freshly harvested or disposed of as waste — into fuels which are similar to those currently in use, such as methane or ethanol [1]. Studies on these topics are carried out in several universities as well as in private and governmental institutes.

1 Methane Production

Methane is produced everywhere in nature where organic compounds are degraded by microorganisms in the absence of oxygen, sulfate and nitrate; thus under these conditions CO_2 is the only available electron acceptor [2]. Methane together with CO_2 (biogas) is formed for example in the lower sediments of rivers or lakes where settled organic matter under oxygen conditions undergoes anaerobic decomposition (marsh gas production). Thus this anaerobic process forms an essential link in the carbon cycle in nature. Partial methane formation occurs also in the gastro-intestinal tract of ruminants; one cow produces about 100–200 l methane per day [3]. It is estimated that up to 10^{12} m^3 of methane are formed by microorganisms each year [4].

On an industrial scale, microbial methane formation is used since the beginning of this century for the stabilization of sewage sludge from municipal wastewater treatment plants [5]. The purpose has mainly been to improve the handling properties of the sludge, so that it could be concentrated to a form suitable for incineration or mechanical disposal. The methane gas produced in these anaerobic digestion processes can be used for the operation of the wastewater treatment plants, and in well designed plants a respectable amount of surplus energy remains for other consumers.

The process of anaerobic digestion has also been used for the treatment of highly concentrated organic liquid animals wastes [6]. In 1947, a special conference was held at Ludwigsburg (near Stuttgart) to discuss the subject of farm digesters [7]. As a result of this meeting, a large mechanized digester with a capacity of 158 m^3 was set up in the Lüneburger Heide [8]. This Bihugas plant, as it was called, became fairly popular, partly because of mechanization, which thus reduced labor requirements, and partly because of high gas production rates. The sizes varied up to 960 m^3 capacity and at the larger installations CO_2 was removed and the methane was stored at a pressure of 350 kp cm^{-2} for use as fuel for tractors [9].

A number of different designs were also developed including for example the horizontal plug flow digester designed by Reinhold and Noack in Darmstadt, Germany, in the early 1950's [6]. This digester received waste of high total solids content which was moved along the covered concrete or steel canal by means of agitators operated several times per day. The contents were heated by steam injection. However, this work was considered at low priority, as during the past 20 to 30 years oil was very cheap. Since the oil crisis in 1973 many Germans have directed their attention once more to anaerobic digestion. Today, research work in several institutes is now concerned with the microbiology and technology of this process to improve its economics.

Present understanding of bacterial populations in anaerobic digesters is rather limited and is based on analysis of bacteria isolated from sewage sludge digesters or from the rumen of some animals. At present it is assumed that three different groups of bacteria are involved in the degradation of organic material into methane and CO_2 [10].

1) Hydrolytic and acid forming bacteria which first hydrolyse the polymers in the biomass (starch, cellulose, proteins, lipids) by extracellular enzymes into their components sugar, amino acids, glycerin and fatty acids. Subsequently these monomers are utilized by the microorganisms and degraded for example into acetic acid, propionic acid, butyric acid, hydrogen, carbon dioxide and various alcohols.

2) The acetogenic bacteria convert the most organic acids to acetate, carbon dioxide and hydrogen. Since these microorganisms can only grow at extremely low partial pressure of hydrogen, this group is largely unknown.

3) The methanogenic bacteria use the metabolites of the acetogenic strains as substrates and produce methane. These bacteria are very strict anaerobes and require a redox potential of about −300 mV. The methanogens represent a phylogenetically unique group of bacteria, which contains special coenzymes and no muramic acid in their cell wall [11, 12].

Thus, although the main groups occuring in the microbial population responsible for anaerobic digestion may be identified, it is also vitally important to understand more fully the biochemical relationships that exist between these populations. The very complex series of transformations that occurs when natural polymers are degraded to methane, carbon dioxide and water is studied in more detail so that control of the process can be optimized for increased efficiency.

The methane production as a function of temperature shows two maxima: one at about 35–40 °C, which corresponds to gas production by mesophilic microorganism; and one at 55–60 °C, corresponding to thermophilic bacteria [2]. There is some evidence that in the thermophilic temperature range, a higher metabolic rate occurs and therefore a higher rate of gas production [13]. Further advantages of operating in this range may be a better destruction of pathogenic organisms and an increased digestion efficiency.

However, since the maintenance of thermophilic conditions requires a higher energy input, the advantages and disadvantages of this process is studied in more detail. In addition to temperature, a number of other process parameters as for example: nutrients, toxic metals, feed concentration and retention time are examined to ensure a proper and safe operation of the digester and for optimization of the methane production rate.

Digesters vary widely with regard to complexity and layout. Many factors affect their arrangement and construction, and these have to be considered in order to optimize their function under each particular set of circumstances and environmental conditions. In Germany there is a potential market for relatively simple as well as for more complicated digesters. Since the disposal of manures from farms with many animals often poses a problem, there is a great interest in inexpensive low technology farm-size digesters [14]. The energy contained in the gas produced would be very welcome on most farms and may be extremely useful in providing for a local shortfall in energy supplies. The residues from the digesters can be used as nitrogen containing fertilizer without high pollution risks. Research work is carried out to develop units which are relatively cheap and easy to operate and maintain. Furthermore the retention time should be shortened to decrease the volume of the digester (probably to the range of 50–100 m^3), which means lower investment and heating costs.

In Germany, the types of digesters used for the sewage treatment are usually more sophisticated. In order to ascertain an optimal process the digester must be probably designed to provide good mixing, constant temperature and pH [15]. Furthermore it is necessary that the system is optimized with regard to waste material input, withdrawal of effluent and collection of gas produced. These digesters usually have capacities between 500 and 700 m^3; the largest European digestion tanks, each with a reaction volume of 12,000 m^3, have been set up at Düsseldorf in 1975 [16]. The maximum gas yield of several German sewage digesters was determined to be about 500 l gas per kg of volatile solids, although in the majority of plants, only values between 100 and 400 l gas per kg volatile solids were obtained. However, it should be mentioned that until now these sewage sludge digesters were not primarily used for gas production but for stabilizing the sludge.

Since recent developments have made a marked reduction in the retention time for the anaerobic fermentation process possible, this process has gained increasing interest in the treatment of highly concentrated industrial wastewater [17]. The anaerobic digestion has two unique advantages over the aerobic biological treatment systems:

1) No energy is necessary for aeration

2) The organic pollutants are converted almost quantitatively to a high energy fuel (biogas) and only negligible excess of microbial biomass (sludge) is formed.

Since some of the anaerobic bacteria have a very long generation time (several days) [10], special reactors have been developed to retain most of the microorganisms inside the digester or to recycle the organism after separation (anaerobic filter, up-flow anaerobic sludge blanket process, anaerobic contact process) [17]. In this way the solid retention time is uncoupled from the fluid retention time, and high bacteria concentrations are obtained in the digester which gives high degradation rates (about 15 kg m^{-3} d^{-1} organic materials). Therefore, it seems almost certain that in the near future large scale digesters will also find their place in those industries producing large quantities of organic waste (breweries, food processing factories, etc.).

Whether or not methane gas derived from anaerobic digesters will play an important role in energy production in West Germany depends very much on the cost of the fuel and the availability of natural organic material. Potential sources of biomass for methane generation are either agricultural, domestic or industrial wastes with a high water content, or plant-crops particular grown for this purpose. However, for a small and densely populated country as Germany, the area which could be made available for energy plantations, is very limited and the use of anaerobic digesters will probably lie mainly in the disposal of wastes. The availability of wastes in West Germany per year is as follows:

1) Municipal refuse: 20×10^6 t containing about: 7.5×10^6 t organic solids

2) Municipal and industrial sewage sludge: 46×10^6 t containing about: 1.4×10^6 t organic solids

3) Animal wastes: 200×10^6 t containing about 18×10^6 t organic solids.

Forest residues and other lignocellulosic waste materials can be only degraded by the anaerobic bacteria after a physical or chemical pretreatment. Furthermore not all of these different organic waste materials are suitable for methane production because of the high costs for collection and transportation. This means that only about 17.5×10^6 t organic solids/year could be made available for anaerobic digestion in West Germany [18]. Assuming that 500 m^3 biogas can be produced from one ton of organic solids, and that 1 m^3 of this gas has an energy content of 27 MJ, the total amount of energy, which could be produced from this organic material, would then be:

$$17.5 \times 10^6 \times 500 \times 27 \text{ MJ} = 236 \times 10^9 \text{ MJ} = 8.1 \times 10^6 \text{ t pit-coal units} .$$

If this amount of energy is put in relation to the total amount of energy consumed in 1979 in West Germany (412×10^7 pit-coal units), it is evident that only about 2% of the total energy requirement can be obtained by this process. However, it has to be mentioned that these calculations have been made on the basis of the energy which can be produced from these wastes by the methane process. For a netto energy balance, the amount of energy which is necessary for collecting and transporting the organic material as well as for the process has to be deducted from these data. For example, about 40–60% of the biogas produced in the sewage sludge digesters is necessary for heating the plants.

2 Ethanol Production

Ethanol can be manufactured by two processes: a) direct hydration of ethylene, or b) degradation of sugars by yeasts. Alcohol formation is used for the production of alcoholic beverages such as beer or wine for several thousands of years. For the production of alcohol used for technical purposes, fermentation has been superceded by the cheaper chemical process in Germany since the thirties. However, the ethanol manufactured for human consumption is required by law to be made by a bioprocess.

In 1973 West Germany produced 80,000 metric tons of fermented alcohol and 117,000 metric tons of synthetic ethanol [19]. However, due to the increasing costs for petroleum the price of synthetic ethanol from ethylene is now nearly equal to that of ethanol from fermentation processes [20]. Therefore, Germany has recently started intensive research and development work on optimizing the alcoholic process.

Ethanol is formed from hexose sugars by yeasts under anaerobic conditions according to the following equation:

$$\underset{(100\ kg)}{C_6H_{12}O_6} \xrightarrow[-O_2]{\text{yeast cells}} \underset{(51\ kg)}{2CH_3CH_2OH} + \underset{(49\ kg)}{2CO_2}$$

Therefore hexose sugars from sugar beet or sugar cane can be used directly for this process, while the polysaccharides as starch from wheat, potatoes, cassava, maize or rice and cellulose from plant residues have to be hydrolyzed before they are accessible to the yeast. The degradation of starch to monosaccharides can be performed by different microbial enzymes. However, in the usual starch mashes a saccharification equilibrium is obtained after about 20 minutes, due to product inhibition of the enzymes. Research work is therefore being carried out with the aim of producing enzymes with lower product inhibition. Since the conventional batch-wise process of liquefication and degradation of starch consumes a lot of energy (about 400 MJ/100 kg starch), a continuous process has been developed with a reduced energy demand of about 40 MJ per 100 kg starch [21].

The enzymatic hydrolysis of cellulose in plant residues is seriously inhibited by lignin, which, due to its close association with the cellulose fibres, acts as a physical barrier. Furthermore cellulose in plant material has both crystalline and amorphous structures and the microbial cellulose degradation is also inhibited by this cristalline structure of cellulose. Therefore, before cellulose can be efficiently hydrolyzed by enzymes to glucose, an inexpensive method which does not increase the waste disposal problem must be found to overcome these natural barriers. Among a number of physical and chemical processes with the potential for enhancing enzymatic degradation of cellulose which have been tested, steaming appears to be a promising method for the pretreatment of wood and crop residues [22]. In order to achieve an economic process for the enzymatic hydrolysis of cellulose into glucose, further work is also carried out to obtain enzymes with higher activity and stability.

The fermentation processes for ethanol production used till now on a commercial scale are batch processes. In recent years, the continuous production of ethanol has recieved worldwide great attention because it offers the advantages of uniform production in large-scale industrial plants, simpler control and a possible increase in productivity. However, the main obstacles are infections decreasing the yield and reduction of fermentation rates by the alcohol formed. Recently, a new continuous process has been developed in Germany [23]. The main constituent of the process is an adapted yeast strain which is characterized by its resistance to infections and ability to grow at elevated temperatures (30–35 °C). In order to increase the stability and fermentative activity of the yeast cells, the culture is fed with a nutrient solution consisting of a mixture of minerals and a small amount of air. As the production rate increases parallel to the cell concentration some of the yeast cells are recycled. Due to the flocculating properties of the yeast used, the cells can be separated from the medium by sedimentation. The alcohol productivity of this process reaches a rate of 50 ml ethanol per 1 volume and h. This rate is about ten times higher than that obtained in other commercial-scale processes. Since higher reaction temperatures and higher ethanol concentrations would be desirable not only with respect to lower infection but also to higher fermentation rates, a screening program has been started to select thermotolerant and ethanoltolerant yeast strains.

As the liquid product generally contains about 6–12% ethanol, large amounts of thermal energy are required for distillation (about 250 kg of steam per 100 l alcohol). Recently, it has been possible to reduce the energy demand in this process to one half by heat exchange and a distillation plant working with several steps at different pressures[24]. Some calculations have shown that the biogas production from anaerobic digestion of distiller's wastes could be sufficient to supply the processes of fermentation and distillation with energy. Furthermore, several experiments have been started to separate the ethanol during fermentation by semipermeable membranes or extraction. The solvent for the extraction should have nearly no solubility for water but a very good solubility for ethanol. The netto energy conversion rate for the production of ethanol from starch could be increased from about 20% in the conventional distillery to at least 70% if all the energy saving steps were included[19]. Whether or not production of ethanol from biomass results in a positive netto energy balance has been the subject of considerable discussions. These results demonstrate that a positive energy balance for ethanol production can be achieved by the use of biomass waste and by the development of novel engineering processes for ethanol recovery.

The use of ethanol as a fuel source for the internal combustion engine is not a new idea. In Europe, during World War II, ethanol-gasoline mixtures were common, but it has been considered economically incompetitive with petroleum products. However, this situation is likely to change soon if the price of petroleum continues to rise sharply. Furthermore ethanol is a non-polluting antiknock fuel. Research is also undertaken to develop engines which do not require anhydrous ethanol but which could run on ethanol with some water content. Need for capital, production and energy costs would be considerably reduced if this was possible. If the oil crisis becomes more severe and ethylene is in short supply, we may revert to alcohol from renewable resources as a basis for the chemical industry. In this case ethanol would be the raw material for ethylene production.

From a study recently carried out in Germany the raw material which could be made available for ethanol production and the use of ethanol as a substitute for gasoline has been calculated[25]. In Germany there are about 22×10^6 cars, lorries and busses which need about 53×10^9 l gasoline per year i.e. 18% of our total energy consumption. Since the heating value of ethanol is significantly lower than that of gasoline about 65×10^9 l ethanol per year would be necessary for substituting the total gasoline requirement. Taking into account the area which is necessary for food production only about 2.7×10^6 ha land could be made available for energy plantations in Germany. From the biomass of this area the total ethanol production would amount to 6.5×19^9 l per year. This corresponds to approximately 10% of the fuel which would be necessary for all the motor cars in Germany. From these data it becomes clear, that ethanol production from biomass can make only a small contribution to our gasoline requirement. However, the addition of about 10% ethanol to nonleaded gasoline would increase the antiknock value of the fuel. Since the costs for the biomass (sugar, starch) and for the plant construction with about 60–80% of the total costs are very high, the production costs of ethanol from biomass are about 1.70–2.30 DM per l, which is approximately four times the momentary price for gasoline. However, this situation is likely to change soon if the price of petroleum continues to increase in the same way as it has in the last few years, and the production of ethanol from agricultural and forest residues is developed.

3 Conclusion

It can be seen that in Germany the microbial methane production from waste organic matter can only deliver a very small part of our total energy consumption, although in rural areas its importance may increase in the near future. The main use of this process will be the treatment of wastes for pollution control, giving methane as a useful by-product. Thorough microbiological and engineering studies are necessary to investigate all the factors which will produce a simple, cheap digester able to operate with a minimum retention time but with maximum energy yield.

Furthermore, for a small and densely populated country as Germany, the biomass wich could be produced for a gasohol production is very limited. At present the greatest challenge is to develop bioprocesses that employ the cheap and underutilized biomass residues, because they are often considered as wastes and require net energy input for treatment without product gain. In the future new developments may be possible to produce more biomass for ethanol production. Undoubtedly, more research on both the fundamentals and applications of the alcohol process is needed before its utility in biofuel and chemical production can be economically assessed.

4 References

1. BMFT-Leistungsplan 04, Biotechnologie, Bundeministerium Forsch. Technolog. 1980
2. Zeikus, J. G.: Bacteriol. Rev. *41*, 514 1977)
3. Wolfe, R. S.: Adv. Microbial Physiol. *6*, 107 (1971)
4. Ehhalt, D. H.: The atmospheric cycle of methane. In: Microbial production and utilization of gases (ed. Schlegel, H. G. et al.), p. 13. Göttingen: E. Goltze 1976
5. Hobson, P. N. et al.: Crit. Rev. Environ. Control. *4*, 131 (1974)
6. Noack, W.: Biogas in d. Landwirtschaft, Darmstadt, Elsner 1955
7. Rosenberg, G.: J. Min. Agric. *58*, 487 (1952)
8. Rosenberg, G.: Farm Mechanisation *3*, 425 (1951)
9. Schmidt, F. et al.: Gas J. *279*, 4757 (1954)
10. Bryant, M. P.: Animal Sci. *48*, 193 (1979)
11. Thauer, R. K. et al.: Naturwissenschaften *66*, 89 (1979)
12. Kandler, O.: ibid. *66*, 95 (1979)
13. Pfeffer, J. T.: Biotech. Bioeng. *16*, 771 (1974)
14. Baader, W. et al.: Biogas in Theorie u. Praxis, Darmstadt, KTBL-Schrift 229, 1978
15. Roediger, H.: Die anaerobe alkalische Schlammfaulung, München, Oldenbourg 1967
16. Loll, U.: Engineering, operation and economies of biodigesters. In: Microbial energy conversion (ed. Schlegel, H. G. et al.), p. 361, Göttingen, E. Goltze 1976

17. Lettinga, G. et al.: Biotech. Bioeng. *22*, 699 (1980)
18. Umweltbundesamt-Bericht: Beitrag der Biogaserzeugung zur Deckung des Primärenergiebedarfs der Bundesrepublik Deutschland, Berlin 1980
19. Dellweg, H. et al.: Dechema Monogr. *83*, 35 (1979)
20. Cheremisinoff, N. et al.: Biomass, applications, technology and production. New York, Marcel Dekker 1980
21. Misselhorn, K.: Chem. Rundschau *38*, 1 (1980)
22. Dietrichs, H. H. et al.: Holzforsch. *32*, 193 (1978)
23. Faust, U. et al.: Kontinuierliche Äthanolherst. durch ein Gärverfahren der Hoechst/Uhde-Biotechnologie. In: 4. Symp.: Tech. Mikrobiologie (ed. Dellweg, H.), p. 37, Berlin 1979
24. Misselhorn, K.: Branntweinwirt. *6*, 91 (1980)
25. Gieseler et al.: Gärungsalkohol aus Agrarprodukten als Biokraftstoff, Studie der Dornier, 1980

Biomass Conversion in South Africa

Hans Jurgens Potgieter
Department of Microbiology University of the Orange Free State Bloemfontein, 9300, Republic of South Africa

South Africa is using or is investigating the potential of forest biomass, sugar-cane, maize, grain sorghum, cannery and industrial wastes, algae and other agricultural crops to contribute to its total annual energy consumption of 3000×10^6 GJ per annum. These materials can also be utilized for the production of chemicals and food. Several factories already exist and some are in the planning stage. There are National programs for the conversion of biomass which is co-ordinated by the Council for Scientific and Industrial Research.

1 Introduction

The conversion of biomass to fuel (energy) and other chemicals and foodstuffs did not receive serious attention in South Africa during the previous decades. It is realized, however, that energy from renewable resources will ultimately have to make a large contribution to replace energy derived from coal and imported oil.

The annual energy consumption in the Republic of South Africa is in the region *of* 3000×10^6 GJ. Coal is currently supplying approximately 75% of the country's total energy demand. Imported oil contributes another 17% and hydropower 5%. Two 1000 MW nuclear reactor units will be commissioned in 1982 and 1984. The South African Coal, Oil and Gas Corporation (Sasol) will be in a position to supply about 40% of South Africa's liquid fuel requirement when Sasol II and Sasol III are fully commissioned. The fuel is obtained from lowgrade coal. From 1956 to 1971 Sasol motorfuel contained as much as 15% C_2 and C_3 alcohols. Currently 10% ± 2% "motor-alcohol" containing 70% ethanol and other $C_3 - C_6$ alcohols in decreasing amounts is blended on a small scale with premium motorfuel to give a 93 octane specification. The extractable coal reserves have been estimated at $25{,}000 \times 10^6$ t. Crude oil has yet to be discovered.

In view of the current world fuel situation the South African Government has accepted in principle that agricultural products as well as waste may be utilized to extend and/or replace petroleum fuels [1].

Municipal, industrial and agricultural waste offer another source of energy. The annual output of solid household refuse in South Africa is 3 to 6 million tonnes.

2 Forest Biomass

The Forestry Council of South Africa favours the production of methanol from wood. They are currently evaluating the economic feasibility of producing methanol from wood including a conceptual demonstration pilot gasifier plant and a commercial plant design [2]. According to Van Breda [2] present and potential timber plantings on readily available good land can supply 26×10^6 m^3 per annum of wood and waste. Good and marginal land can supply a further 15×10^6 m^3 p.a. It is estimated that there will be a potential surplus of 22×10^6 m^3 or 11×10^6 oven dry metric tonnes p.a. The conversion of this material to methanol or ethanol can contribute a further 25% of the country's liquid fuel needs.

3 Surgar-Cane Waste

Sugar-cane is planted for the production of sugar.

If sugar processing waste is defined as bagasse and molasses it is estimated that at current production rates 15% of South Africa's motorfuel needs could be obtained from this source. Surplus sugar, however, can increase this percentage.

3.1 Molasses

3.1.1 Ethanol and Related Products

The use of molasses as raw material for ethanol production in South Africa dates back more than a hundred years. Three large distilleries (National Chemical Products (Transvaal), National Chemical Products (Natal) and Natal Cane By-Products Ltd.) use molasses as a raw material for ethanol production by *Saccharomyces cerevisiae* [3]. National Chemical Products (Transvaal) originally used maize as a raw material and the ethanol produced was added to motorfuel. The two plants of National Chemical Products (NCP) use the ethanol for the supply of industrial alcohol, potable alcohol, methylated spiritis, and absolute alcohol. Both liquid and solid CO_2 is sold. The ethanol stream is also diverted for the integrated production of acetaldehyde, acetic acid, ethyl acetate, froth flotation reagents etc. Natal Cane By-Products Ltd. is the only local producer of diethyl-ether.

3.1.2 Acetone Butanol

The NCP facility in Germiston also produce acetone, butanol and ethanol (6:3:1 ratio) from molasses using *Clostridium acetobutylicum* [3].

3.1.3 Animal Feed

It is obvious that the effluent from the bioreactors can create a serious effluent problem. NCP, however, convert all the effluent from their plants to "Dried Molasses Distillers Solubles". This is then used as base for the "Rumevite System" of ruminant nutrition which is manufactured and marketed by subsidiaries world wide [3].

3.1.4 Fodder Yeast

Natal Cane By-Products Ltd. channel their vinasse (slop) after distillation to a subsidiary for the production of fodder yeast [3]. After the addition of molasses a continuous process is employed, the material centrifuged, dried and sold for animal nutrition. A small quantity goes into human foods.

3.1.5 Yeast

Aerobic conversion of cane molasses by various strains of *Saccharomyces cerevisiae* is used by four companies for the production of active dried yeast, active dried sorghum beer yeast, active dried yeast and compressed bakers yeast. The total annual output is approximately 20,000 t [3].

3.2 Bagasse

The Council for Scientific and Industrial Research (CSIR) in Pretoria co-ordinates a national programme for the conversion of cellulose and hemi-cellulose from bagasse to liquid fuel and other products. The total annual yield of cane crushed in the 1975–1976 season was 16,813,530 t [4]. This production was increased to about 19×10^6 t in 1978. The fibre mass fraction is 0.1567, therefore the total fibre associated with the 1978 crop was 2.97×10^6 t. The average chemical composition of this fraction amounts to 43% cellulose, 35% pentosans and 22% lignin [5]. Land areas as yet unexploited for cane production can increase the fibre yield with 1.25×10^6 t per year.

3.2.1 National Programme Participants

The following laboratories in South Africa participate in the bagasse programme and their specific activities are given.

National Food Research Institute, C.S.I.R., Pretoria

a) Microbiology Research Group:
   Isolation and improvement of aerobic cellulolytic organisms.

b) Fermentation Laboratory:
   Process technology for the large scale production of enzyme broth.

Sugar Milling Research Institute, University of Natal, Durban

Pre-treatment of bagasse to enhance enzymatic hydrolysis.

University of the Orange Free State, Bloemfontein

a) Department of Biochemistry:
   Enzymatic conversion of pentoses to fermentable substrates.
b) Department of Microbiology:
   i) Physiological parameters for the enzymatic saccharification of acid extracted bagasse and fermentation to ethanol.
   ii) Isolation and improvement of organisms capable of converting hemicellulose hydrolysates to ethanol and 2,3-butane diol.

University of Fort Hare, Alice

Department of Biochemistry:
The cellulase complex of *Trichoderma reesei.*

University of Cape Town, Cape Town

a) Department of Chemical Engineering:
   i) Fermentation kinetics of microorganisms producing ethanol and other solvents from xylose and other pentoses derived from hemicellulose.
   ii) Ethanol fermentation from bagasse hydrolysate.
b) Department of Microbiology:
   i) Production of solvents from hemicelluloses and pentoses from bagasse by *Clostridium acetobutylicum.*
   ii) Genetic studies on *Clostridium thermocellum* to provide improved microorganisms for simultaneous saccharification/fermentation of sugar-cane bagasse to ethanol.
c) Department of Chemistry:
   Chemical analysis of bagasse and molecular weight distribution measurements of breakdown products.

University of Natal, Durban

Department of Chemical Engineering:
Acquisition of chemical and biochemical data for the design of a continuous or semi-continuous process to convert bagasse to xylose and glucose.

University of Natal, Pietermaritzburg

Department of Microbiology:
i) Isolation and improvement of aerobic and anaerobic microorganisms for the utilization of bagasse following acid pretreatment.
ii) Saccharification of cellulose and hemicellulose by aerobic and anaerobic microorganisms.

University of Durban-Westville, Westville

Department of Microbiology:
Evaluation and minimization of cellulase/yeast antagonism in simultaneous saccharification/fermentation.

4 Pineapple Waste

Pineapple canneries in the East London area must dispose of approximately 50,000 t pineapple peel and 50,000 t "concentrated" pineapple effluent annually which is equivalent to 7000 t waste carbohydrates. The "concentrated" pineapple effluent is collected from various discharge points before being mixed and diluted with waste water from other parts of the cannery plant. The Department of Microbiology at the University of the Orange Free State Bloemfontein [6,7] showed that the total effluent can be converted to produce 3.5% ethanol. This is equivalent to 2×10^6 l ethanol. By supplementing every ton of peel/effluent mixture with 0.13 t molasses the ethanol concentration can be increased to 8% (v/v) ethanol within a 72 h process period which would facilitate economic distillation. More than 8×10^6 l of ethanol could be produced in this manner annually. The solids will be processed for animal nutrition. A low-technology facility for processing the pineapple peel/effluent mixture is designed and erection of the plant will probably start in 1981.

5 Maize

The total maize crop harvested each year is in the vicinity of 11×10^6 t. An equivalent tonnage of plant residues containing mainly cellulose, hemi-cellose and lignin is produced. Part of the grain crop can be utilized for ethanol production. The Sentrachem group of companies has standard modular designs for bioreactors producing either 50,000, 100,000 or 150,000 m^3 of ethanol p.a. using either maize or grain sorghum as substrate. A 100,000 m^3 plant will require approximately 250,000 t of maize grain per year. The building of these facilities are awaiting government approval.

6 Grain Sorghum

Grain sorghum has great potential and can also be grown on good and marginal land. The bioreactors mentioned above for maize is so constructed that it can process grain sorghum which is currently produced in sufficient quantities to meet the local demand.

7 Sunflower

The agricultural and transport sector of the economy depends highly on diesel fuelled movers [8]. Suitable substitutes for diesel or diesel fuel extenders must be sought. The

Republic of South Africa produces between 400,000 and 500,000 t of sunflower seed p.a., mainly for cooking oil, margarine, and animal feed. This yield can be bettered by agricultural crop manipulation. The on-farm input of liquid fuel to produce 600 l of sunflower seed oil is 60 l.

The Division of Agricultural Engineering of the Department of Agriculture and Fisheries is persuing exciting ideas [1,8] on the utilization of a sunflower seed oil based diesel substitute. By changing pure sunflower seed oil to an ethyl or a methyl ester mixture the viscosity and other physical properties of the sunflower seed oil are changed. A typical mixture consists of 60% by volume of fatty acid ethyl esters, 25% unreacted sunflower seed oil, and 15% ethyl alcohol. Results so far obtained are excellent and in certain aspects the mixture even performs better than diesel.

8 Algae

Algae mass culture is unlikely to produce the large quantities of cheap proteïn and energy initially expected. Several laboratories in South Africa have programmes on algal culture for the production of animal or human food, and specific by-products.

AECI at Modderfontein [9] has a commercial 10,000 m^2 algal pond utilizing a weak factory effluent made up of boiler water blow-downs, cooling water purges and plant washwater and stormwater with a nitrogen content of 100 $mg\, l^{-1}$ N which otherwise will be a pollution hasard.

The Institute of Environmental Sciences at the University of the Orange Free State, Bloemfontein has several projects on open [10] and closed [11] algal cultures. In the closed system $12 \times 2\ m^2$ miniponds and one 100 m^2 pond are available for experimental purposes. These experiments are run in collaboration with Sentrachem with the aim of combining mass algal culture systems with conventional ethanol units.

9 Plants as Sources of Natural Rubber

The Council for Scientific and Industrial Research also coordinates National Programmes on the potential of *Parthenium argentatum* Gray (Guayule), *Simmondsia chinensis* Link (Jojoba) and other promising plants like *Llandolphia kirkii* and *Manihot glaziovii* as sources of natural rubber.

10 References

1. Bruwer, J. J. et al.: 1980 Symp. S. Afr. Inst. Agr. Eng. Pretoria (1980)
2. Van Breda, P. V.: S. A. Food Rev. *7*, 133 (1980)
3. Lurie, J.: S. A. Food Rev. *7*, 136 (1980)
4. Lamusse, J. P.: Proc. 50th Congr. S. Afr. Sugar Technologists Assoc. p. 149, 1976
5. Dekker, R. F. H., Lindner, W. A.: S. Afr. J. Sci. *75*, 65 (1979)
6. Prior, B. A. et al.: S. A. Food Rev. *7*, 120 (1980)
7. du Preez, J. C., Prior, B. A., Lategan, P. M.: Symp. on Aquaculture in Wastewater. Paper No. 7 C.S.I.R. Pretoria (1980)

8. Bruwer, J. J. et al.: Amer. Şoc. Agr. Eng. Energy Symp. Kansas City. Mo. Oct. 1, 1980
9. Bosman, J., Hendricks, F.: Symp. on Aquaculture in Wastewater. Paper No. 2 C.S.I.R., Pretoria 1980
10. Grobbelaar, J. U.: ibid. Paper No. 1
11. Walmsley, R. D., Shilfinglaw, S. N., Cronje, L.: ibid. Paper No. 28

Biomass Utilization in Switzerland

Th. Haltmeier
Swiss Federal Institute of Technology Hönggerberg, 8093 Zürich

1 Introduction

In Switzerland the research and development programs for the bioconversion of biomass to energy, chemicals and single cell protein (SCP) are in their infancy although this field will achieve greater importance due to the expected shortage of fossil fuels. Even so, the estimated share of bioenergy will be relatively small compared to the whole energy demand. The two main sources of bioenergy in Switzerland will be wood and biogas: The maximal exploitation of our forest and the use of all wood waste- and byproducts will yield 2.85% of the whole energy demand [1]. From agricultural sources, if manure and the relevant volumes of waste materials such as straw and corn stalks were used for methane production, 1.6% of the whole energy demand could be generated by this source [2]. To obtain these two maximal values in biomass utilization one has to overcome several barriers: for maximal exploitation of our forests, the mountainous region has to be opened up; problems involved in such an attempt include: lack of roads, impassable areas and an insufficient labour force for the task. The achievement of maximal biogas-production is hindered by the lack of cheap technology for storage and utilization of the gas during the summer. Many farms are too small for economic gas production. Agricultural waste-products occur in relatively small amounts in a decentralized distribution. They are used to a large extent as animal foodstuffs, fertilizers and to strew in cowsheds. This situation is illustrated by the fact that between the years 1971–1976 80'000 t of straw had to be imported annually [3]. The use of energy-crops and energy-farming is discussed but no studies are undertaken for reasons of political and economic considerations. In Switzerland the agriculturally useful area per inhabitant is small and at the present time 35% of the actual calorie demand of the country is imported [4]. The fact that biomass is a regenerable resource of raw-materials and energy emphasizes its importance. The research activities are briefly described in the following pages.

2 Photosynthesis

Basic research in the transformation of sunlight to chemical energy takes place at the Universities of Bern and Neuchâtel, whilst the Friedrich-Miescher-Institute in Basle studies the possibility of raising the yield of photosynthesis in corn. The various institutes and research centres of agriculture have an old tradition in looking at plant genetics. The production of hydrogen using algae is investigated by a group at the University of Zurich.

3 Wood

In Switzerland the following priority for the utilization of wood exists: timber for construction, timber for the cellulose and paper industries and lastly for energetic use. Actually different studies are carried out which investigate the occurrence, distribution and application of timber and wood waste-material. In the mountain village of Sent (Unterengadin, GR) the NEFF (Nationaler Energie Forschungs Fond) supports a pilot plant heating the whole village. The wood of the community will be burnt and the gas produced is used to run a heat pump supported by suncollectors. Hydrolysis of wood for the production of sugars, ethanol and SCP, respectively, is performed neither on an industrial scale or examined in pilot plants. Investigation in this field is restricted to feasibility studies. Ethanol is produced from spent sulfite liquor by the Cellulose-Attisholz Company. The energy conception for the near future indicates an increase for the share of wood of the total energy demand from 1.2% (1978) to 2.85% (2000). This represents a volume of 3 million m^3 of wood. 40% of this amount will be taken from the forests, the remainder being composed of all kinds of wood waste materials [1]. Wood will still be burnt in stoves or for central heating, but other technologies such as gasification together with the coupling of heat and power in bigger units will probably be applied in the future [5].

4 Agricultural and Municipal Waste-Materials

The research program "Nationales Forschungsprogramm 7B" (NFP7B) supporting studies in the field of biogenic raw- and waste-material, is sponsoring two studies to record the extent of usable waste-materials. One study is engaged in examining all forms of waste-material and their viability. The basis for management policy is considered by looking at different technologies, the economic situation, and the present legal position, in order to promote applications which are efficient from an overall economic point of view. Another study investigates the flow of energy and waste material on farms.

Applied and basic research takes place in several institutes and research centers. Agricultural research centers are examining the influence of sewage sludge and municipal waste compost on the fertility of soil by its use as fertilizer. EAWAG and ETH-Zürich are running a project sponsored by the NFP7B which examines the possibility of an anaerobic thermophilic activated sludge treatment to get a hygienic and stable sludge for agricultural use. The Batelle Geneva Research Centres have been involved in a wide range of technical and economic feasibility studies associated with biomass conversion into fuels, chemicals, fertilizers or SCP. The Batelle Research Centre developed a pilot plant for the acid hydrolysis of vegetable residues to sugars and their subsequent microbial conversion to alcohol and SCP [6]. A research program at the ETH-Zürich examines the influence of different pretreatment methods on the biological degradation of wheat straw [7].

5 Biogas

There are about 30 agricultural biogas plants in Switzerland of which half are demonstration or experimentation plants [8]. The gas produced is mainly used for heating purposes and in two cases electricity is also produced. At the agricultural research center "FAT" a study investigating the significance and the feasibility of the use of biogas on farms was made [2]. In addition, examinations and measurements on existing gasplants are done and new plants for research purposes are under construction at several agricultural research centres. A private research team sponsored by NEFF started to explore the kinetics of the methane-production. A group of researchers at the ETH-Zürich are engaged in a project to make a cheese dairy self supporting in energy. Pigs kept at this dairy are fed on whey and the methane produced from the waste of these pigs helps supply the energy demand. At a waste-water treatment plant in Altenrhein attempts are being made to increase yield of methane in a pilot plant, the influence of thermophilic conditions is also examined. Research in the microbial conversion of methane to methanol is done by a private company. At the "FAT" a project investigating the feasibility of the use of biogas as fuel in agricultural machines has just started.

6 SCP

A group of scientists at the ETH-Zürich is investigating the possibility of producing a protein-rich foodstuff for animals from agricultural wastes such as straw and whey, using fungi and yeast. This project is part of the COST-program 83/84. The Batelle Research Center investigates a system where sludge is converted to SCP by means of protozoa [9].

7 Abbreviations

COST	European Cooperation in the field of Scientific and Technical Research
EAWAG	Eidgenössische Anstalt für Wasserversorgung, Abwasserreinigung und Gewässerschutz
ETH	Eidgenössische Technische Hochschule
FAT	Eidgenössische Forschungsanstalt für Betriebswissenschaft und Landtechnik
NEFF	Nationaler Energie Forschungs-Fond
NFP	Nationales Forschungs-Programm, Schweizerischer Nationalfonds zur Förderung der wissenschaftl. Forschung

8 References

1. Bundesamt für Forstwesen: Wald- u. Holzwirtschaft in d. Energiepolitik. 1980*
2. Kaufmann, R.: Systemstudie über möglichen Umfang und Bedeutung der Biogaserzeugung u. Verwertung in d. Landwirtschaft. Schriftenreihe der FAT (in press)
3. Schweizerisches Bauernsekretariat. Statist. Erheb. u. Schätzungen über Landwirtschaft u. Ernährung. 1977
4. Studer, R.: Landwirtschaft u. Energiepolitik*
5. Hegetschweiler, Th.: Problemkatalog für Forschung u. Entwicklung im Bereich Holzenergie*
6. Batelle Geneva Res. Center: Biotechnology, Nov. 1979
7. Binder, A., Haltmeier, Th., Fiechter, A.: Pretreatment of straw. II. Int. Symp. Bioconversion and Biotech. Eng., IIT Delhi 1980 (in press)
8. Göbel, W.: Landwirtschaftl. Biogasanlagen in d. Schweiz, Übersicht. Biogastagung 8.—10. May, FAT 1979
9. Ayerley, A.: Enzyme Microb. Technol. *2*, 54 (1980)

* papers presented at: Informationstagung über Biomassen-Verwertung zur Energiegewinnung. FAT, 2. Mai 1980

Swedish Developments in Biotechnology Based on Lignocellulosic Materials

K.-E. Eriksson
Swedish Forest Products Research Laboratory, Box 5604, S-114 86 Stockholm, Sweden

This article describes the basic research carried out in Sweden on enzyme mechanisms involved in the fungal degradation of cellulose and lignin. It also depicts the biotechnical processes based on lignocellulosic materials that have been developed or are under development in this country. A description of Swedish biomass resources is also given.

1 Introduction

One of nature's most important biological process is the degradation of lignocellulosic materials to carbon dioxide, water and humic substances. This conversion is catalysed by enzymes produced by microorganisms. An understanding of these reactions has been shown to be important for the modification and improvement of existing forest industrial processes and for the substitution of chemical processes by biological ones.

Biotechnologies with different aims can be developed by utilizing microorganisms and their enzymes for the conversions of biomass. The biomass may come from forestry or from agriculture. One type of biotechnology that has become particularly interesting in these days of increasing oil prices and shortage of fossil fuel resources is the conversion of biomass into energy-rich chemicals and fuel-ethanol.

We describe the research — both basic and applied — which has been carried out on the bioconversion of lignocellulosic materials in Sweden. In some areas, this research has only started recently and the projects can therefore only be briefly indicated.

The possibility of creating biotechnological processes based on lignocellulosic materials depends, to a large degree, on an understanding of how these materials are converted by microorganisms on the molecular level. A considerable amount of this knowledge has been obtained through basic research into these processes in Sweden.

2 Swedish Biomass Resources

The domestic raw materials which can be used for the production of chemicals and fuels in Sweden can be generated from the growth of energy forests, from forestry waste in general and from other biomasses.

Energy forests can be characterized as areas cultivated with trees particularly chosen for fast growth. These trees can be hardwood species such as salix and populus. Species of this kind will be cultivated where soil conditions, water and nutrient conditions, etc. are favourable. The selected hardwood species will, after plantation, need a cultivation time of 1–3 years before the first harvest. The idea is to harvest in cycles of 1–3 years. Each energy forest growth is expected to have a favourable life-length of 20–30 years before a new plantation is necessary.

The trees from the energy forests will be harvested with the aid of special, newly developed and highly effective machine systems. Machine equipment will vary, depending both on the size of the growth and also on factors such as the carrying power of the soil, etc.

The collected fibre mass must be fractionated, dried, stored and transported before use. If the biomass is burned, the residue product formed will be mainly ash. For certain applications, reuse of the nutrient salts in the ashes will be possible.

In a historical perspective, forest energy (mainly residual products such as branches, tops and small-sized wood) has been utilized for large-scale energy production in Sweden. This is particularly true of the period between the use of coal and oil in the country. The total wood resources in Sweden are estimated to be 2.3 billion m^3. Under air-dried conditions (25% water), this is equivalent to 500 million tons of oil.

In Sweden, about 84 million m^3 of wood are cut each year, of which 10 million m^3 are not utilized. Theoretical estimation of the annual amount of waste in Swedish forest point to a figure of about 46 million m^3 solid measure. Part of this waste is expected to be used in the future to meet the demands of the pulp industry, but the total amount of forest waste is estimated to be equivalent to 3–5 million tons of oil.

Several systems for the collection and preparation of this energy resource are expected for the future. These can be developed parallel to the development of other techniques for large-scale production for use within the pulp industry. A new line of development is the cutting and collection of whole trees which are processed (debranched, debarked, cut and chipped) in a central terminal. Another line is to sort the biomass before it leaves the forest into different categories with different final uses.

Other biomasses can be divided into waste materials and into biomasses specially cultivated as an energy source. Among the waste biomasses, the greatest interest is projected towards straw but reed is also of interest. In addition, attention is being devoted to the waste from households and industries, and also to another waste material from agriculture, namely manure.

Examples of plants cultivated for energy production, in addition to the energy forests already mentioned, include certain herbaceous plants with a very fast growth. In addition, the large-scale production of seaweed and algae can be promoted in certain artificial environments.

3 Basic Research

3.1 Enzyme Mechanisms Involved in Fungal Cellulose Degradation

The utilization of biomass in biotechnological processes implies, as stated in the introduction, a thorough knowledge of the enzymatic reactions involved in the degradation and conversion of these materials.

In Sweden, the enzyme mechanisms involved in cellulose degradation have been extensively studied for two fungi, namely the white-rot fungus *Sporotrichum pulverulentum* [1)] and the mould *Trichoderma reesei* [2)].

For the fungi *S. pulverulentum* and *T. reesei*, the pattern of attack on cellulose may be summarized as follows: The fungus *S. pulverulentum* hydrolyses cellulose through the action of: a) five different endo-1,4-β-glucanases which attack at random 1,4-β-linkages along the cellulose chain; b) one exo-1,4-β-glucanase which splits off cellobiose or glucose units from the non-reducing end of the cellulose; c) two 1,4-β-glucosidases which hydrolyse cellobiose and water-soluble cellodextrins to glucose and cellobionic acid or glucose and glucono lactone [3–6)]. It has been generally accepted that essentially the same picture is also valid for cellulose hydrolysis by *T. reesei* [7)]. However, a few differences have been recognized including the number of the various hydrolytic enzymes and the degree to which the β-glucosidase activity is bound to the fungal cell wall.

In *S. pulverulentum*, an oxidative enzyme important for cellulose degradation has been discovered in addition to the hydrolytic enzymes described above [8)]. The enzyme has been purified and characterized and found to be a cellobiose oxidase, which oxidizes cellobiose and higher cellodextrins to their corresponding onic acids thereby using molecular oxygen. The enzyme is a hemoprotein and also contains a FAD group. It is not yet known whether this enzyme also oxidizes the reducing endgroup formed in cellulose when 1,4-β-glucosidic bonds are split through the action of endo-glucanases.

Another unconventional enzyme produced for the conversion of cellobiose by *S. pulverulentum* is the enzyme cellobiose: quinone oxidoreductase [9–11)]. This enzyme is important for the degradation of both cellulose and lignin. Although the enzyme seems to be involved in both lignin and cellulose degradation, the highest enzyme production was reached when cellulose powder was used as a carbon source.

The oxidoreductase is relatively specific of its disaccharide substrates, but the specificity for quinone substrates is much less pronounced. The enzyme is able to reduce both ortho- and para-quinones. The total reaction scheme for cellulose degradation in *S. pulverulentum* is presented in Fig. 1.

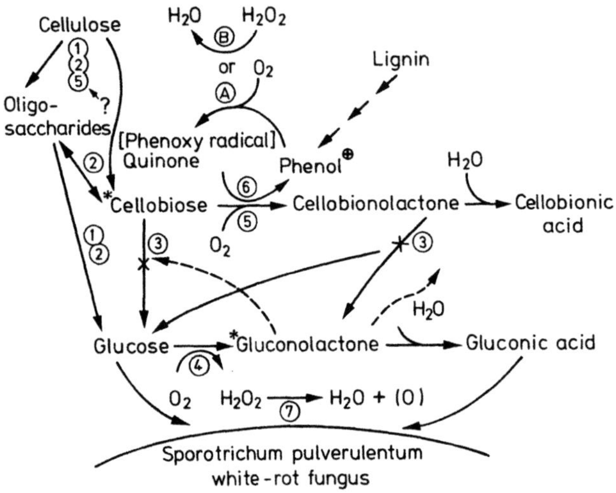

* Products regulating enzyme activity; gluconolactone [3], cellobiose increase transglycosylations
$^+$ Products regulating enzyme synthesis; glucose, gluconic acid » catabolite repression, » phenols repression of glucanases

Fig. 1. Enzyme mechanisms for cellulose degradation and their extra-cellular regulation in *Sporotrichum pulverulentum*. From Eriksson [34]. The following enzymes are involved in the reactions:
1. endo-1,4-β-glucanases
2. exo-1,4-β-glucanases
3. 1,4-β-glucosidase
4. glucose oxidase
5. cellobiose oxidase
6. cellobiose:quinone oxidoreductase
7. catalase

Enzymes involved in lignin degradation
A. laccase
B. peroxidase

Not only are the enzymes participating in cellulose degradation by *S. pulverulentum* known but the regulatory mechanisms involved in this degradation are also understood. The fact that glucono lactone is a very powerful inhibitor of the β-glucosidases produced by *S. pulverulentum* has been evidenced by recent studies in our laboratory. The extracellular-1,4-β-glucosidase activity in *S. pulverulentum* can be split into two main peaks. The k_i-values for glucono lactone inhibition in the two β-glucosidases are 3.5×10^{-7} och 15×10^{-7} M, respectively. The corresponding k_i-value for *T. reesei* QM 9414 1,4-β-glucosidase was found to be 3.2×10^{-5} M.

The regulation of endo-1,4-β-glucanase production in *S. pulverulentum* has recently been investigated using a newly developed sensitive method [12]. The results show that cellobiose causes induction of endo-1,4-β-glucanases at concentrations as low as

1 mg l^{-1}. It was also shown that glucose causes catabolite repression of enzyme formation at concentrations as low as 50 mg l^{-1}. Mixtures of inducer and repressor give rise to a delayed enzyme production compared with solutions of inducer alone.

Studies of the mould *T. reesei* QM6A using the same technique show that cellobiose under our conditions is not an inducer of endo-1,4-β-glucanases. However, sophorose causes induction of endo-1,4-β-glucanases at a concentration of 1 mg l^{-1}, as has recently been confirmed by studies of Sternberg and Mandels [13]. The comparison between the regulation of endo-1,4-β-glucanase production in the two fungi also demonstrates several other important differences. For example, a solution of CMC alone induces enzyme formation in *S. pulverulentum* but not in the *T. reesei* strain. Under our experimental conditions, no 1,4-β-glucanases were actively excreted into the solution by *T. reesei*. This has previously been reported also by Berg and Pettersson [14]. Although they used cellulose as a carbon source, the enzymes were bound to the cell wall. However, it has recently been shown [15,13] that sophorose gives rise to active excretion of endo-1,4-β-glucanases into the cultures of *T. reesei* QM9414.

At the Biomedical Center, University of Uppsala, Sweden, Dr. Pettersson and coworkers have for a long time been concerned with studies of cellulases from *Trichoderma* strains. The goal has been to determine how cellulose is enzymatically degraded. Their studies have included investigations into the mechanisms for the individual enzymes involved in the process and into how these enzymes cooperate. The project in Uppsala has been mainly basic research. However, the general aim of the studies has been to obtain results that would be of as great value as possible for applied research.

The first studies by Pettersson's group were carried out with a commercial Japanese product, Onozuka SS, as an enzyme source. From this material two endo-glucanases, one exo-glucanase and one β-glucosidase were purified and characterized. These results are listed in Table 1 [2].

It was soon found that the composition of the Onozuka SS material varied considerably. In some batches, for instance, a certain enzyme could be completely missing. For this reason, *T. reesei* QM 9414 was used instead as an enzyme source.

Table 1. Some properties of cellulolytic enzymes isolated from *Trichoderma viride* [2]

Type of enzyme	Activity towards different substrates						
	Molecular weight	Isoelectric point	Carbohydrate content (%)	CMC	Microcrystalline cellulose	Repricipitated cellulose	Cellotetraose
Exo-1,4-β-glucanase	42000	3.79	9	−	+	+	+
Endo-1,4-β-glucanase I	12500	4.60	21	+	−	+	+
Endo-1,4-β-glucanase II	50000	3.39	12	+	−	+	+
β-Glucosidase	47000	5.74	0	−	−	−	+

From culture solutions four different cellulolytic enzymes have been purified. Two of these are endo-glucanases of the same type as characterized earlier, one is a cellobiohydrolase and one is of a type which has not earlier been characterized. However, the latter seems to have properties similar to those of the exo-glucanase described in Table 1. The enzymes are summarized in Table 2, the new exo-glucanase-type enzyme being designed as C_0. The culture medium from *Trichoderma* QM 9414 also contains a low β-glucosidase activity. Preliminary experiments have demonstrated that a synthetic mixture of endo-glucanase H, cellobiohydrolase and the C_0-enzyme is as effective as the original culture medium when cellulose undergoes degradation, i.e. if small amounts of a β-glucosidase are added. As in earlier studies, a strong synergistic effect between the endo-glucanases and the cellobiohydrolase was found. A pronounced synergism has also been found between cellobiohydrolase and C_0-enzyme. Since both the C_0-enzyme and the cellobiohydrolase enzyme seem to be exo-enzymes, it is difficult to explain this newly discovered exo-exo synergism.

Table 2. Some properties of cellulolytic enzymes isolated from *Trichoderma reesei* QM 9414 (Pettersson et al. [2])

Type of enzyme	Molecular weight	Isoelectric point	Number of isocomponents
Endo-glucanase L	20000	7.5	1
Endo-glucanase H	51000	4.7	4
Cellobiohydrolase	42000	3.9	4
C_0	~50000	5.8	3

In cooperation with Dr. M. Mandels at the Army Natick Laboratories, Mass., USA and Dr. Bland S. Montenecourt, Rutgers University N.J., USA, Dr. Pettersson's group has studied the enzyme production for some mutants of *T. reesei* QM 9414. By immunochemistry it has been possible to demonstrate that one of the mutants forms approximately twice as much of the cellobiohydrolase enzyme as the wild-type. QM 9414, cultivated under optimal conditions, produces 3.4 g cellobiohydrolase and 0.8 g endo-glucanase H/l of culture solution. This mutant is thus a fantastic producer of cellulases.

In recent years, amino acid-sequence determination of the cellobiohydrolase enzyme was started by the Uppsala group. At present, this determination is nearly terminated. Sequence work has now been started on the endo-glucanase H enzyme. To understand better the relations between structure and function of these enzymes, X-ray crystallographic studies have been started in cooperation with Birkbeck College, University of London, England.

3.2 Degradation of Lignin by the White-Rot Fungus *Sporotrichum pulverulentum*

Lignin is a phenyl-propanoid polymer of vascular plants which gives the plants rigidity and binds plant cells together [16]. Lignin also decreases water permeation across cell walls of xylem tissue and protects plant tissues from invasion by pathogenic organisms. This very complex polymer is not readily attacked by microorganisms. The most successful group of organisms in lignin degradation, the white-rot fungi, are still the only microorganisms which have been shown to be capable of totally degrading all the major wood components.

It is generally accepted that lignin biosynthesis is catalysed by phenol oxidases [17]. The involvement of phenol oxidases in the degradation of lignin has also been discussed ever since Bavendamm [18] used gallic and tannic acid to differentiate between white-rot and brown-rot fungi. The reason for this interest in phenol oxidases may be that white-rot fungi, which utilize and degrade lignin, also produce extra-cellular phenol oxidases in contrast to brown-rot fungi. Furthermore, lignin contains phenolic units which constitute a substrate for phenol oxidases.

In view of the above, it seemed natural to study the importance of phenol oxidases in lignin degradation [19]. For these studies, three different strains of *S. pulverulentum* were utilized namely a) the wild-type; b) a phenol oxidase-less mutant, Phe 3 (obtained by UV-irradiation of wild-type spores), and c) a phenol oxidase-positive revertant, Rev 9 (obtained by UV-irradiation of spores from Phe 3). The phenol oxidase-less mutant did not degrade lignin or any other wood component. The revertant, however, degraded all wood components, including lignin, to the same extent as did the wild-type strain. After addition of purified laccase to kraft lignin agar plates, the phenol oxidase-less mutant could again degrade lignin almost as well as the wild-type, indicating that only the gene controlling the synthesis of phenol oxidase was affected by the mutagenic treatment. These results point to an obligatory role of phenol oxidases in lignin degradation.

It has been demonstrated in several studies that vanillic acid is always a metabolic product of lignin degradation by white-rot fungi [20]. Our efforts to use different isolated lignins as substrates for the submerged growth of *S. pulverulentum*, in order to evaluate the enzyme mechanisms involved in the degradation of lignin, failed, partly due to the difficulty of achieving conditions suitable for degradation. We therefore decided to use vanillic acid as substrate. The strategy has been to approach the problem "enzyme mechanisms involved in lignin degradation by white-rot fungi" by working initially with small molecules, lignin models, and approaching the lignin polymer as such using more and more complex substances.

The results of the studies of vanillic acid degradation by *S. pulverulentum* are presented in Fig. 2. Figure 2 shows that vanillic acid is simultaneously oxidatively decarboxylated to methoxyhydroquinone (MHQ) and reduced to vanillin and vanillyl alcohol. The decarboxylation pathway is more predominant in shake cultures whereas reduction predominates in standing cultures. The reduction steps also seem to require energy in the form of an externally supplied, easily metabolized carbon source such as glucose or cellobiose.

Vanillate seems to be metabolized inside the fungal cell. The intracellular vanillate hydroxylase (catalysing decarboxylation) has been isolated and purified and some of its characteristics have been described by Buswell et al. [21]. The phenol oxidases laccase and peroxidase can also decarboxylate vanillic acid but are independent of NAD(P)H for their activity [22].

It has also been demonstrated that *S. pulverulentum* produces an aromatic ring-cleaving enzyme [23]. This enzyme does not cleave the ring unless the latter bears three hydroxy groups. By using differently labelled vanillic acids it was possible

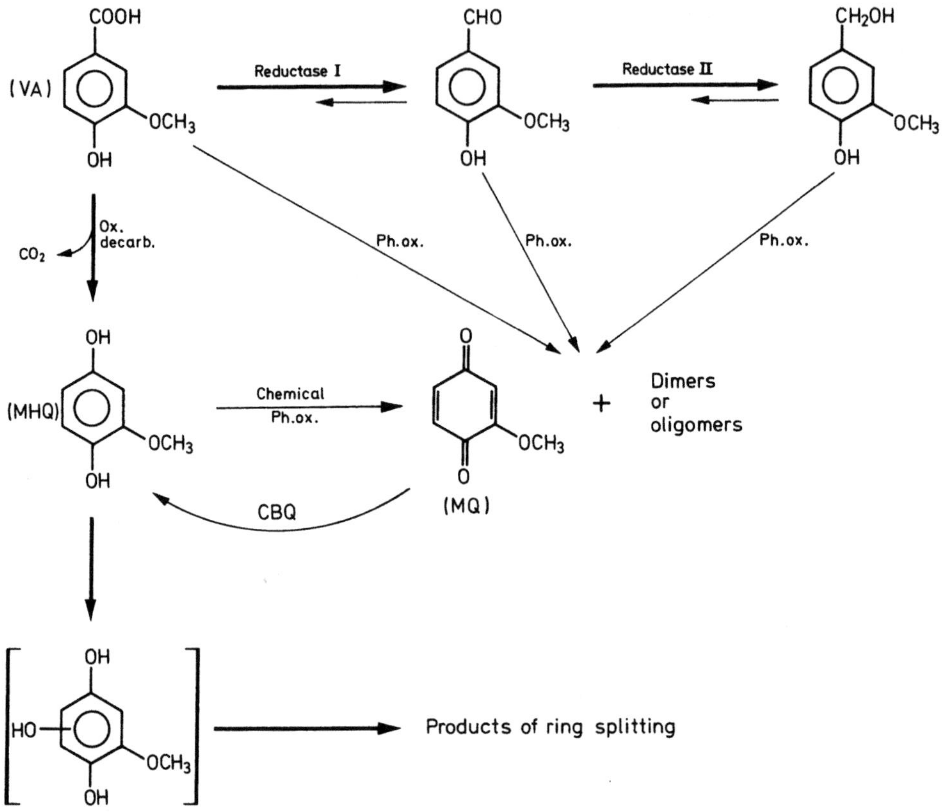

Fig. 2. Proposed routes for the metabolism of vanillic acid by *S. pulverulentum*. OH—MHQ is a hydroxylated MHQ which may be the true substrate for ring cleavage

to show that decarboxylation takes place before the ring is cleaved which, in turn, occurs earlier than the release of $^{14}CO_2$ from $O^{14}CH_3$-vanillate. The results obtained suggest that a third hydroxy group is introduced into the ring via direct hydroxylation rather than via demethylation, which does not appear to take place until after the ring is cleaved [22].

Quinones are readily formed by the action of phenol oxidases induced during the growth of white-rot fungi on both low molecular weight phenolic compounds and lignin. Although certain quinones are normal components of cellular electron transport systems, quinones are generally highly reactive and are known to inhibit a wide range of metabolic processes. Therefore, reduction of quinoid intermediates is apparently essential. We have earlier reported that extracellular reduction of quinones and phenoxy radicals take place via the extracellular enzyme cellobiose:quinone oxidoreductase [9–11]. This enzyme makes use of cellobiose as a co-substrate. Cellobiose is thereby oxidized to the corresponding lactone. In studying the vanillic acid metabolism, we have now found a second, intracellular quinone oxidoreductase system from *S. pulverulentum* which reduces quinones to hydroquinones using pyridine nucleotides as electron donors [24].

We are now concerned with the development of growth systems that will allow us to study more efficiently the degradation of the lignin polymer as such. The technique will involve the use of labelled lignin and of a solid-state growth system. We expect that such a growth system will allow a characterization of the enzymes involved in the attack on the lignin polymer and the degradation products formed. We have already been able to separate the degradation products from lignin by the aid of HPLC. The characterization of the fragmentation products is in progress.

4 Biotechnological Processes

4.1 A Process Serving as a "Kidney" in Closed White Water Systems of Forest Product Industries

Mechanical pulp can be produced from wood by three different methods, a) stone grinding, b) refining in a refiner and c) so-called thermomechanical pulping (TMP) in which wood chips are refined under high temperature and pressure. Mechanical pulp is used for newsprint papers, various grades of board, etc. The various treatments to produce mechanical pulp give rise to soluble substances, i.e. sugars, lignins, etc. in the white-water systems, particularly if the TMP process is used. The sugars are a mixture of monomers and water-soluble oligomers, all excellent substrates for fungi that produce extracellular enzymes for hydrolysing these polymers.

Two processes for the production of fodder protein based on waste materials from forest products industries are already in technical use and may serve as examples of existing biotechnology in the pulp and paper industry. These processes are the *Candida utilis* process and the Pekilo process based on the fungus *Paecilomyces varioti.* In both these processes the substrate is mainly the monosaccharides in spent sulfite liquor. Disaccharides and higher oligosaccharides are utilized only to a very limited extent.

The organic substances dissolved in mechanical pulping, monomeric and oligomeric sugars, phenols, etc., are not very good substrates for either the Candida or the Pekilo process. In a big newsprint mill, approximately 25–40 t of water-soluble sugars or phenols are produced per day. It is clear that this waste water must be purified before it can be released into lakes and streams. Both chemical and biological purification is necessary to treat the effluents from mechanical pulping which requires large-scale investments.

In our laboratory we have developed a process using the dissolved substances from mechanical pulping as substrates [25]. This process is based on the white-rot fungus *S. pulverulentum.* This fungus can readily utilize dissolved substances in waste-fibre building board liquors. The residence time for continuous cultivation is approximately 17 h with this substrate. The same residence time was found for the degradation of the white water from a newsprint paper mill when the process was recently tested on a pilot plant scale. With this residence time no build-up of organic matter took place.

A considerable reduction in residence time is achieved if some of the mycelium produced in the continuous process is recycled to the fermentor. The approach is being tested at present. Recycling of mycelium is necessary to keep down the fermentor volume. During the process the fungal mycelium is obtained in pellet form. Since the size of these pellets is about 0.2 to 0.4 mm they can easily be separated from the culture solution by filtration.

The process has also been run on a 25 m^3-scale using waste water from a fibre-building board factory as substrate. The protein produced, i.e. the fungus *S. pulverulentum*, has been tested in feeding trials at The Swedish University of Agricultural Sciences, Uppsala. These feeding trials are essentially positive, particularly with ruminants [26]. However, the digestibility of the fungal cell wall by mono-gastric animals is not as good as the digestibility by ruminants.

The production of fodder protein by our process is economical only if the costs of alternative water purification are also taken into account. However, the fungal mycelium produced need not be used as cattle feed; it can also be added to the paper. This possibility has been investigated and the paper properties are not significantly influenced by an addition of fungal mycelium corresponding to 1.5% of the paper weight. The benefits of closing a paper mill system by the application of our process can be listed as follows; 1) water purification, prevention of the build-up of organic matter in the white-water system eliminating the need for external water purification; 2) a higher process temperature is obtained in the pulping section by closing the white-water system; the hot water can be converted into steam to be used for drying the paper; 3) increased paper production by addition of the fungus to the paper; 4) water savings; by closing the system only approximately 10% of the water necessary in an open system is used.

4.2 Biomechanical Pulping

Mechanical pulping is a process that consumes much of the electrical energy produced in Sweden. Different ways of reducing the energy consumption in mechanical pulping have therefore been tried. Our approach has been to use cellulase-poor mutants of white-rot fungi to remove some of the lignin from wood chips [27]. We have demonstrated that such a decrease in the lignin content leads to energy saving in mechanical pulping. It must be stressed, however, that our investigations of these possibilities have not yet led to an industrial process.

Our approach to the use of microorganisms for the production of mechanical pulp has been as follows: cellulase-poor mutants of white-rot fungi have been obtained by irradiation of fungal spore suspensions with UV-light [27, 28]. The spores are then plated out on cellulose agar plates containing small amounts of glucose to allow the cellulase-less mutants to grow. A chemical, for instance saponin, is also added to the plates in order to obtain colonial growth. After about one week, clearance zones are obtained around most of the colonies where the cellulose has been degraded and solubilized. Colonies without these clearance zones do not degrade cellulose since they no longer excrete the necessary enzymes. We have now available several different mutants from white-rot fungi. The most successful so far has been Cel 44, a cellulase-poor mutant from *S. pulverulentum* [27]. The optimal growth conditions in wood for the mutant as well as for the wild-types have been investigated [29,30]. To understand better how the wood components are degraded by the fungi on the molecular level and to study the changes affected in the wood cell walls by fungal pretreatment, we found it important to investigate the morphology of the fungal attack on the wood fibre. Such studies have been undertaken both on the micromorphological level and on the ultrastructural level using scanning electron microscopy (SEM) and transmission electron microscopy (TEM) [31,32]. By SEM it has been found that the cellulase-less mutants penetrate the wood mainly through existing channels and holes. The wild-type fungi are independent of these passages and bore holes straight through the fibre cell walls, thereby damaging the fibres.

By transmission electron microscopy if has been demonstrated that enzymes attacking the lignin diffuse from the fungal hypha although only short distances. The application of this technique has also revealed that pretreatment of wood with cellulase-less mutants gives rise to cellulose fibrils which are more visible than those in untreated wood.

Pulp and paper production from wood chips pretreated with cellulase-less mutants allows the following conclusions to be drawn: Rotting leads to

- a more rapid decrease in freeness (i.e. less energy is required to reach a certain freeness level),
- a higher density (at a certain input of energy),
- a lower tear index (at a certain input of energy),
- a lower tensile index (at a certain density),
- a lower light scattering and a higher light absorption.

The most interesting relationship from an energy-saving point of view is possibly that between tensile index and energy input. So far, the results are similar to those for untreated wood. Both positive and negative results have been obtained. However, recent results indicate that when the chips to be rotted are impregnated with small amounts of sugar, the tensile index versus energy input relationship is influenced in a positive way. An approximate energy saving of 20% is obtained.

In addition to the possibility of using cellulase-less mutants for the pretreatment of wood chips in order to save energy in mechanical pulping, it is also possible to use such organisms to upgrade for instance straw and sugar-cane bagasse as cattle feed. Inoculation of these materials with a mutant will lead to delignification with increased digestibility of the delignified material by the cattle.

5 Ethanol Production Based on Lignocellulosic Materials

The increasing pressure upon the fossil fuel resources has aroused world-wide interest in the production of fuels and chemicals from renewable resources. One area which is particularly considered at present is the production of ethanol via sugar from lignocellulosic materials [33].

Supported by government money and in cooperation with a major Swedish company, we are at present trying to develop a Swedish process for ethanol production based on biomass. There are several obvious advantages with such an ethanol production in Sweden. One of these, and maybe the most important, is that fuel can be produced from our own resources. In the event of an embargo on oil imports, methods for the domestic production of fuel will be vitally important. Another advantage is that we can cut down on our import of fossil fuels and thus positively influence the balance of trade.

Ethanol production based on biomass involves many process steps, such as delignification, saccharification, ethanol formation and distillation. The project has recently started and is still in an investigative stage. Decisions with regard to which lines of development we shall work on will be made later this year.

6 References

1. Eriksson, K.-E.: Pure Appl. Chem. *53*, 33 (1981)
2. Pettersson, L. G.: Symposium on Enzymatic Hydrolysis of Cellulose. Bailey, M., Enari, T.-M., Linko, M. (eds.), p. 255, Helsinki 1975
3. Eriksson, K.-E., Pettersson, B.: Eur. J. Biochem. *51*, 193 (1975)
4. Eriksson, K.-E., Pettersson, B.: Eur. J. Biochem. *51*, 213 (1975)
5. Streamer, M., Eriksson, K.-E., Pettersson, B.: Eur. J. Biochem. *59*, 607 (1975)
6. Deshpande, V., Eriksson, K.-E., Pettersson, B.: Eur. J. Biochem. *90*, 191 (1978)
7. Ryu, D. D. Y., Mandels, M.: Enzyme Microb. Technol. *2*, 91 (1980)
8. Ayers, A. R., Ayers, S. B., Eriksson, K.-E.: Eur. J. Biochem. *90*, 171 (1978)
9. Westermark, U., Eriksson, K.-E.: Acta Chem. Scand. *B 28*, 204 (1974)
10. Westermark, U., Eriksson, K.-E.: Acta Chem. Scand. *B 28*, 209 (1974)
11. Westermark, U., Eriksson, K.-E.: Acta Chem. Scand. *B 29*, 419 (1975)
12. Eriksson, K.-E., Hamp, S. G.: Eur. J. Biochem. *90*, 183 (1978)
13. Sternberg, D., Mandels, M.: J. Bacteriol. *139*, 761 (1979)
14. Berg, B., Pettersson, G.: J. Appl. Bacteriol.: *42*, 65 (1977)
15. Gritzali, M., Brown, Jr. R. D.: Adv. Chem. Ser. *181*, 237 (1979)
16. Sarkanen, K. V., Ludwig, C. H., in: Lignins: occurrence, formation, structure and reactions. Sarkanen, K. V., Ludwig, C. H. (eds.), p. 43. Wiley-Interscience, New York 1971
17. Freudenberg, K., Neish, A. C.: Constitution and Biosynthesis of Lignin, Springer Verlag, Berlin-Heidelberg-New York 1968
18. Bavendamm, W.: Z. Pflanzenkr. *38*, 257 (1928)
19. Ander, P., Eriksson, K.-E.: Arch. Microbiol. *109*, 1 (1976)
20. Kirk, T. K., Connors, W. J., Zeikus, J. G.: The structure, biosynthesis, and degradation of wood, in: Recent Adv. Phytochem. 11. Loewus, F. A., Remeckles, V. C. (eds.), p. 369, Plenum Press, New York 1977
21. Buswell, J. A. et al.: FEBS Lett. *103*, 98 (1979)
22. Ander, P., Hatakka, A., Eriksson, K.-E.: Arch. Microbiol. *125*, 189 (1980)
23. Buswell, J. A., Eriksson, K.-E.: FEBS Lett. *104*, 258 (1979)
24. Buswell, J. A., Hamp, S. G., Eriksson, K.-E.: FEBS Lett. *108*, 229 (1979)
25. Ek, M., Eriksson, K.-E.: Biotech. Bioeng. *22*, 2273 (1980)
26. Thomke, S., Rundgren, M.: Biotech. Bioeng. *22*, 2285 (1980)
27. Ander, P., Eriksson, K.-E.: Svensk Papperstidn. *78*, 643 (1975)
28. Eriksson, K.-E., Goodell, E. W.: Can. J. Microbiol. *20*, 371 (1974)
29. Eriksson, K.-E., Vallander, L.: Biomechanical pulping, in: Microbiology, Chemistry and Applications. Kirk, T. K., Higuchi, T., Chang, H.-M. (eds.), Boca Raton: CRC Press Inc. 1980
30. Eriksson, K.-E., Grünewald, A., Vallander, L.: Biotech. Bioeng. *22*, 363 (1980)
31. Eriksson, K.-E., Grünewald, A., Nilsson, T., Vallander, L.: Holzforschung *34*, 207 (1980)
32. Ruel, K., Barnoud, F., Eriksson, K.-E.: Holzforschung (in press)
33. Edemar, L.-G., Eriksson, K.-E.: Svensk Papperstidn. *10*, 271 (1980)
34. Eriksson, K.-E.: Biotech. Bioeng. *20*, 317 (1978)

Author Index Vol. 1–20

H. Brauer, Y. B. G. Varma

Air Pollution Control Equipment

1981. 285 figures, 53 tables.
VII, 388 pages.
ISBN 3-540-10463-1

Air Pollution Control Equipment is a critical compilation of available knowledge in the design and operation of equipment for abating harmful emissions. Air pollution control technology, though of relatively recent origin, has already achieved a high degree of efficacy due to the intense research and development invested in it in past decades.

This book is addressed primarily to chemical, environmental, and mechanical engineers who are engaged in the design and operation of equipment for air pollution control. But it will certainly be helpful to chemists and physicists who are confronted with the solutions of environmental problems. Furthermore it is intended as a textbook for engineering courses on environmental protection.

Springer-Verlag
Berlin
Heidelberg
New York

GPSR Compliance

The European Union's (EU) General Product Safety Regulation (GPSR) is a set of rules that requires consumer products to be safe and our obligations to ensure this.

If you have any concerns about our products, you can contact us on ProductSafety@springernature.com

In case Publisher is established outside the EU, the EU authorized representative is:

Springer Nature Customer Service Center GmbH
Europaplatz 3
69115 Heidelberg, Germany

Batch number: 10620881

Printed by Printforce, the Netherlands